INTERNATIONAL ENERG

BIOFUELS
FOR
TRANSPORT

An International Perspective

FOREWORD

The IEA last published a book on biofuels in 1994 (*Biofuels*). Many developments have occurred in the past decade, though policy objectives remain similar: improving energy security and curbing greenhouse gas emissions are, perhaps more than ever, important priorities for IEA countries. And, more than ever, transportation energy use plays a central role in these issues. New approaches are needed to cost-effectively move transportation away from its persistent dependence on oil and onto a more sustainable track. But technology has made interesting progress and this will continue in the coming years, creating new opportunities for achieving these objectives.

It is not surprising that interest in biofuels – and biofuels production – has increased dramatically in this past decade. Global fuel ethanol production doubled between 1990 and 2003, and may double again by 2010. In some regions, especially Europe, biodiesel fuel use has also increased substantially in recent years. Perhaps most importantly, countries all around the world are now looking seriously at increasing production and use of biofuels, and many have put policies in place to ensure that such an increase occurs.

This book takes a global perspective in assessing how far we have come – and where we seem to be going – with biofuels use in transport. It reviews recent research and experience in a number of areas: potential biofuels impacts on petroleum use and greenhouse gas emissions; current and likely future costs of biofuels; fuel compatibility with vehicles; air quality and other environmental impacts; and recent policy activity around the world. It also provides an assessment of just how much biofuels could be produced in OECD and non-OECD regions, given land requirements and availability, what the costs and benefits of this production might be, and how we can maximise those benefits over the next ten years and beyond.

Claude Mandil
Executive Director, IEA

ACKNOWLEDGEMENTS

This publication is the product of an IEA study undertaken by the Office of Energy Efficiency, Technology and R&D under the direction of Marianne Haug, and supervised by Carmen Difiglio, Head of the Energy Technology Policy Division. The study was co-ordinated by Lew Fulton and Tom Howes. The book was co-authored by Lew Fulton, Tom Howes and Jeffrey Hardy. Additional support was provided by Rick Sellers and the Renewable Energy Unit.

The IEA would like to express its appreciation to David Rodgers and John Ferrell of the US Department of Energy for their advice and support in developing the analysis that led to this publication. The IEA would also like to acknowledge the following individuals who provided important contributions: Jean Cadu (Shell, UK); Christian Delahoulière (consultant, Paris); Mark Delucchi (U. C. Davis, US); Thomas Gameson (Abengoa Bioenergía, Spain); Mark Hammonds (BP, UK); Francis Johnson (Stockholm Environment Institute, Sweden); Luiz Otavio Laydner (CFA Banco Pactual, Brazil); Lee Lynd (Dartmouth College, US); Kyriakos Maniatis (EU-DG-TREN, Brussels); Tien Nguyen (US DOE, US); Isaias de Carvalho Macedo (Centro de Tecnologia Copersucar, Brazil); Jose Roberto Moreira (Cenbio, Brazil); Suzana Kahn-Ribeiro (COPPE/UFRJ, Brazil); Bernhard Schlamadinger (Joanneum Research, Austria); Harald Schneider (Shell, Germany); Leo Schrattenholzer (IIASA, Austria); Ralph Sims (Massey U., NZ); Don Stevens (Pacific Northwest National Laboratory, US); Björn Telenius (National Energy Admin., Sweden); Marie Walsh (Oak Ridge National Laboratory, US); Michael Quanlu Wang (Argonne National Laboratory, US); and Nick Wilkinson (BP, UK).

Assistance with editing and preparation of the manuscript was provided by Teresa Malyshev, Muriel Custodio, Corinne Hayworth, Bertrand Sadin and Viviane Consoli.

TABLE OF CONTENTS

FOREWORD 3

ACKNOWLEDGEMENTS 4

EXECUTIVE SUMMARY 11

1 INTRODUCTION 25

What are Biofuels? 26

Global Biofuel Production and Consumption 27

2 FEEDSTOCK AND PROCESS TECHNOLOGIES 33

Biodiesel Production 33

Ethanol Production 34

Biomass Gasification and Related Pathways 43

Achieving Higher Yields: the Role of Genetic Engineering 47

3 OIL DISPLACEMENT AND GREENHOUSE GAS REDUCTION IMPACTS 51

Ethanol from Grains 52

Ethanol from Sugar Beets 57

Ethanol from Sugar Cane in Brazil 57

Ethanol from Cellulosic Feedstock 61

Biodiesel from Fatty Acid Methyl Esters 63

Other Advanced Biofuels Processes 64

4 BIOFUEL COSTS AND MARKET IMPACTS 67

Biofuels Production Costs 68

Biofuels Distribution and Retailing Costs 86

Biofuels Cost per Tonne of Greenhouse Gas Reduction 91

Crop Market Impacts of Biofuels Production 94

5 VEHICLE PERFORMANCE, POLLUTANT EMISSIONS AND OTHER ENVIRONMENTAL EFFECTS   **101**

Vehicle-Fuel Compatibility   *101*

Impacts of Biofuels on Vehicle Pollutant Emissions   *111*

Other Environmental Impacts: Waste Reduction, Ecosystems, Soils and Rivers   *118*

6 LAND USE AND FEEDSTOCK AVAILABILITY ISSUES   **123**

Biofuels Potential from Conventional Crop Feedstock in the US and the EU   *124*

Ethanol Production Potential from Cellulosic Crops   *133*

Other Potential Sources of Biofuels   *137*

Biofuels Production Potential Worldwide   *138*

7 RECENT BIOFUELS POLICIES AND FUTURE DIRECTIONS   **147**

IEA Countries   *147*

Non-IEA Countries   *157*

Outlook for Biofuels Production through 2020   *167*

8 BENEFITS AND COSTS OF BIOFUELS AND IMPLICATIONS FOR POLICY-MAKING   **171**

The Benefits and Costs of Biofuels   *171*

Policies to Promote Increased Use of Biofuels   *179*

Areas for Further Research   *186*

ABBREVIATIONS AND GLOSSARY   **191**

REFERENCES   **195**

LIST OF TABLES

Table 1.1   World Ethanol Production and Biodiesel Capacity, 2002   30

Table 3.1   Energy and GHG Impacts of Ethanol: Estimates from Corn- and Wheat-to-Ethanol Studies   53

Table 3.2   Net Energy Balance from Corn-to-Ethanol Production: A Comparison of Studies   56

Table 3.3   Estimates from Studies of Ethanol from Sugar Beets   59

Table 3.4   Energy Balance of Sugar Cane to Ethanol in Brazil, 2002   60

Table 3.5   Estimates from Studies of Ethanol from Cellulosic Feedstock   62

Table 3.6   Estimates from Studies of Biodiesel from Oil-seed Crops   63

Table 3.7   Estimates of Energy Use and Greenhouse Gas Emissions from Advanced Biofuels from the Novem/ADL Study, 1999   65

Table 4.1   Estimated Corn-to-Ethanol Costs in the US for Recent Large Plants   69

Table 4.2   Ethanol Cost Estimates for Europe   71

Table 4.3   Engineering Cost Estimates for Bioethanol plants in Germany, and Comparison to US   72

Table 4.4   Ethanol Production Costs in Brazil, circa 1990   75

Table 4.5   Cellulosic Ethanol Plant Cost Estimates   78

Table 4.6   Gasoline and Ethanol: Comparison of Current and Potential Production Costs in North America   79

Table 4.7   Biodiesel Cost Estimates for Europe   80

Table 4.8   Estimates of Production Cost for Advanced Processes   83

Table 4.9   Ethanol Transportation Cost Estimates for the US   88

Table 4.10   Cost of Installing Ethanol Refuelling Equipment at a US Station   90

Table 4.11   Total Transport, Storing, and Dispensing Costs for Ethanol   91

Table 4.12   Estimated Impacts from Increased Use of Biodiesel (Soy Methyl Ester) in the US   96

Table 4.13   Estimated Impacts from Increased Production of Switchgrass for Cellulosic Ethanol on Various Crop Prices   97

Table 5.1   Changes in Emissions when Ethanol is Blended with Conventional Gasoline and RFG   113

Table 5.2   Flexible-fuel Vehicles (E85) and Standard Gasoline Vehicles (RFG): Emissions Comparison from Ohio Study   115

Table 5.3	Biodiesel/Diesel Property Comparison	117
Table 6.1	Biofuel Feedstock Sources	123
Table 6.2	Ethanol and Biodiesel Production: Comparison of US and EU in 2000	125
Table 6.3	Typical Yields by Region and Crop, circa 2002	127
Table 6.4	Biofuels Required to Displace Gasoline or Diesel	129
Table 6.5	US and EU Biofuels Production Scenarios for 2010 and 2020	132
Table 6.6	Estimated Cellulosic Feedstock Availability by Feedstock Price	134
Table 6.7	Post-2010 US Ethanol Production Potential from Dedicated Energy Crops (Cellulosic)	136
Table 6.8	Estimates of Long-term World Biomass and Liquid Biofuels Production Potential	140
Table 6.9	Current and Projected Gasoline and Diesel Consumption	144
Table 6.10	Cane Ethanol Blending: Supply and Demand in 2020	144
Table 7.1	Transportation Fuel Tax Rates in Canada	149
Table 7.2	EU Rates of Excise Duty by Fuel, 2003	151
Table 7.3	Current EU Country Tax Credits for Ethanol	152
Table 7.4	UK Annual Vehicle Excise Duty for Private Vehicles	155
Table 8.1	Potential Benefits and Costs of Biofuels	172

LIST OF FIGURES

Figure 1	Range of Estimated Greenhouse Gas Reductions from Biofuels	13
Figure 2	Biofuels Cost per Tonne of Greenhouse Gas Reduction	16
Figure 1.1	World and Regional Fuel Ethanol Production, 1975-2003	28
Figure 1.2	World and Regional Biodiesel Capacity, 1991-2003	29
Figure 2.1	Ethanol Production Steps by Feedstock and Conversion Technique	35
Figure 4.1	US Ethanol Production Plants by Plant Size, as of 1999	70
Figure 4.2	Average US Ethanol and Corn Prices, 1990-2002	71
Figure 4.3	US and EU Average Crop Prices, 1992-2001	73
Figure 4.4	Prices for Ethanol and Gasoline in Brazil, 2000-2003	77

Figure 4.5   Cost Ranges for Current and Future Ethanol Production   85

Figure 4.6   Cost Ranges for Current and Future Biodiesel Production   86

Figure 4.7   Cost per Tonne of CO_2 Reduction from Biofuels in Varying
Situations   92

Figure 4.8   Biofuels Cost per Tonne GHG Reduction   93

Figure 5.1   Potential Emissions Reductions from Biodiesel Blends   117

Figure 6.1   Estimated Required Crops and Cropland Needed
to Produce Biofuels under 2010/2020 Scenarios   131

Figure 6.2   Cane Ethanol Production, 2020, Different Scenarios   143

Figure 7.1   Fuel Ethanol Production, Projections to 2020   167

Figure 7.2   Biodiesel Production Projections to 2020   169

Figure 8.1   Ethanol Import Duties Around the World   185

LIST OF BOXES

The Biodiesel Production Process   34

The Sugar-to-Ethanol Production Process   36

The Grain-to-Ethanol Production Process   37

The Cellulosic Biomass-to-Ethanol Production Process   40

IEA Research in Bioenergy   43

Hydrogen from Biomass Production Processes   46

The Net Energy Balance of Corn-to-Ethanol Processes   57

Macroeconomic Impacts of Biofuels Production   99

Ethanol and Materials Compatibility   102

Diesel Fuel and the Cetane Number   110

Recent WTO Initiatives Affecting Biofuels   185

EXECUTIVE SUMMARY

Biofuels for transport, including ethanol, biodiesel, and several other liquid and gaseous fuels, have the potential to displace a substantial amount of petroleum around the world over the next few decades, and a clear trend in that direction has begun. This book looks both at recent trends and at the outlook for the future, in terms of potential biofuels production. It also examines the benefits and costs of biofuels use to displace petroleum fuels. It takes an international perspective, assessing regional similarities and differences and recent activities around the world.

Compared to petroleum, the use of biofuels for transport is still quite low in nearly every country. By far the largest production and use is of ethanol in the United States and Brazil, where similar volumes are used – many times higher than in any other country. But even in the United States, ethanol represents less than 2% of transport fuel (while in Brazil it accounts for about 30% of gasoline demand). However, many IEA countries, including the US, Canada, several European countries (and the European Union), Australia and Japan, are considering or have already adopted policies that could result in much higher biofuels use over the next decade. Many non-IEA countries are also adopting policies to promote the use of biofuels.

Biofuels Benefits and Costs

A principal finding is that, while biofuels production costs are fairly easy to measure, the benefits are difficult to quantify. But this does not necessarily mean that the benefits are not substantial. Increasing the use of biofuels can improve energy security, reduce greenhouse gas and pollutant emissions, improve vehicle performance, enhance rural economic development and, under the right circumstances, protect ecosystems and soils. Because these benefits are difficult to quantify, the market price of biofuels does not adequately reflect them. This disadvantages biofuels relative to petroleum fuels. In IEA countries, liquid biofuels production costs currently are high – up to three times the cost of petroleum fuels. But concluding that biofuels are "expensive" ignores the substantial non-market benefits, and the fact that these benefits are increasing as new, more environment-friendly production techniques are developed. In some countries, such as Brazil, biofuels (namely ethanol) production costs are much lower than in IEA countries and are very near the cost of producing petroleum fuel. This will also likely occur in coming years in other countries, as production costs continue to decline.

One important reason why the benefit-cost picture for biofuels is likely to improve in IEA countries in the future is the development of advanced processes to produce biofuels with very low net greenhouse gas emissions. New conversion technologies are under development that make use of lignocellulosic feedstock, either from waste materials or grown as dedicated energy crops on a wide variety of land types. Most current processes rely on just the sugar, starch, or oil-seed parts of few types of crops and rely on fossil energy to convert these to biofuels. As a result, these processes provide "well-to-wheels"[1] greenhouse gas reductions on the order of 20% to 50% compared with petroleum fuels. But new processes under development can convert much more of the plant – including much of the "green", cellulosic parts – to biofuels with very low, possibly zero, net greenhouse gas emissions. The first large-scale cellulose-to-ethanol conversion facility is expected to be built in 2006, most likely in Canada (EESI, 2003). If the cost targets for cellulosic ethanol production techniques over the next decade are met, a new supply of relatively low-cost, high net-benefit biofuels will open, with large resource availability around the world.

In most countries embarking on biofuels initiatives, the recognition of non-market benefits is often the driving force behind efforts to increase their use. These benefits include:

- **Reductions in oil demand.** Biofuels can replace petroleum fuels in today's vehicles. Ethanol is easily blended up to at least 10% with modern conventional gasoline vehicles, and to much higher levels in vehicles that have been modified to accommodate it. Biodiesel can be blended with petroleum diesel fuel in any ratio up to 100% for operation in conventional diesel engines (small amounts of ethanol can also be blended with diesel under certain conditions). Reductions are not, however, 1:1 on a volume basis since biofuels have a lower energy content. Some petroleum is also used to produce biofuels. Our review of "well-to-wheels" studies indicates that it typically takes 0.15 to 0.20 litres of petroleum fuel

---

1. *"Well-to-wheels" refers to the complete chain of fuel production and use, including feedstock production, transport to the refinery, conversion to final fuel, transport to refuelling stations, and final vehicle tailpipe emissions.*

to produce 1 litre of biofuel (with petroleum used to make fertilisers, to power farm equipment, to transport feedstock and to produce final fuels). The use of crops with low fertiliser requirements (such as some grasses and trees) can improve this ratio.

- **Reductions in greenhouse gas emissions.** Ethanol and biodiesel provide significant reductions in greenhouse gas emissions compared to gasoline and diesel fuel on a "well-to-wheels" basis. While a range of estimates exists, Figure 1 shows that most studies reviewed find significant net reductions in CO_2-equivalent emissions for both types of biofuels. More recent studies tend to make estimates towards the higher reduction end of the range, reflecting efficiency improvements over time in both crop production and ethanol conversion. Especially large reductions are estimated for ethanol from sugar cane and from cellulosic feedstocks. Estimates for sugar cane ethanol are based on only two studies, both for Brazil, resulting in the narrow range of estimates.

Figure 1

Range of Estimated Greenhouse Gas Reductions from Biofuels

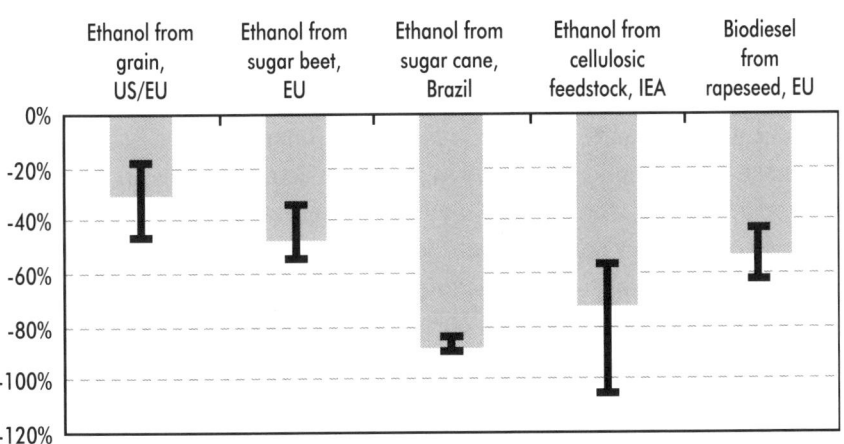

Note: This figure shows reductions in well-to-wheels CO_2-equivalent GHG emissions per kilometre from various biofuel/feedstock combinations, compared to conventional-fuelled vehicles. Ethanol is compared to gasoline vehicles and biodiesel to diesel vehicles. Blends provide proportional reductions; *e.g.* a 10% ethanol blend would provide reductions one-tenth those shown here. Vertical black lines indicate range of estimates; see Chapter 3 for discussion.

- **Air quality benefits and waste reduction.** Biofuels can provide air quality benefits when used either as pure, unblended fuels or, more commonly, when blended with petroleum fuels. Benefits from ethanol and biodiesel blending into petroleum fuels include lower emissions of carbon monoxide (CO), sulphur dioxide (SO_2) and particulate matter (particularly when emissions control systems are poor, such as in some developing countries). Biofuels are generally less toxic than conventional petroleum fuels and in some cases they can reduce wastes through recycling – in particular agricultural wastes from cropland and waste oils and grease that can be converted to biodiesel. However, the use of biofuels can also lead to increases in some categories of emissions, such as evaporative hydrocarbon emissions and aldehyde emissions from the use of ethanol.

- **Vehicle performance benefits.** Ethanol has a very high octane number and can be used to increase the octane of gasoline. It has not traditionally been the first choice for octane enhancement due to its relatively high cost, but with other options increasingly out of favour (leaded fuel is now banned in most countries and methyl-tertiary-butyl-ether [MTBE] is being discouraged or banned in an increasing number of countries), demand for ethanol for this purpose and as an oxygenate is on the rise, *e.g.* in California. In Europe, ethanol is typically converted to ethyl-tertiary-butyl-ether (ETBE) before being blended with gasoline. ETBE provides high octane with lower volatility than ethanol, though typically is only about half renewably derived. Biodiesel can improve diesel lubricity and raise the cetane number, aiding fuel performance.

- **Agricultural benefits.** Production of biofuels from crops such as corn and wheat (for ethanol) and soy and rape (for biodiesel) provides an additional product market for farmers and brings economic benefits to rural communities. But production of biofuels can also draw crops away from other uses (such as food production) and can increase their price. This may translate into higher prices for consumers. The trade-off is complicated by extensive farm subsidies in many countries. These subsidies may in some cases be shifted towards biofuels production, and away from other purposes, as biofuels production rises. In such cases, the net level of subsidy to biofuels production may be much lower than is often assumed.

In contrast to these difficult-to-quantify benefits, the cost of producing biofuels is easier to measure. In IEA countries, the production cost of ethanol and biodiesel is up to three times that of gasoline and diesel. Production costs have dropped somewhat over the past decade and probably will continue to drop, albeit slowly, in the future. But it does not appear likely that biofuels produced from grain and oil-seed feedstock using conventional conversion processes will compete with gasoline and diesel, unless world oil prices rise considerably. Technologies are relatively mature and cost reductions are ultimately limited by the fairly high feedstock (crop) costs. However, the use of lower-cost cellulosic feedstock with advanced conversion technologies could eventually lead to the production of much lower-cost ethanol around the IEA.

The cost story differs in developing countries with sunny, warm climates. In Brazil, feedstock yields of sugar cane per hectare are relatively high; efficient co-generation facilities producing both ethanol and electricity have been developed; and labour costs are relatively low. Thus, the cost of producing ethanol from sugar cane is now very close to the (Brazilian) cost of gasoline on a volumetric basis and is becoming close on an energy basis. The economics in other developing countries, such as India, are also becoming increasingly favourable. As production costs continue to drop with each new conversion facility, the long-term outlook for production of cane ethanol in the developing world appears promising.

Keeping in mind that many benefits of biofuels are not adequately captured in benefit/cost analysis, it is nonetheless important to assess the cost-effectiveness of biofuels for greenhouse gas reduction. Figure 2 compares the cost of reducing greenhouse gas emissions from several types of ethanol. Taking into account just well-to-wheels GHG reductions and incremental costs per litre, in a standard analysis, one can see that ethanol from grain in IEA countries currently costs US\$ 250 or more per tonne of CO_2- equivalent GHG emissions. In contrast, if large-scale plants using advanced conversion processes were constructed today, ethanol from cellulosic feedstocks would cost more per litre, but would provide GHG reductions at a lower cost per tonne (around \$200). Over the next decade the costs of producing cellulosic ethanol may drop considerably, bringing cost per tonne down to \$100 or even \$50. Ethanol produced today in Brazil, with an incremental cost of \$0.03 to \$0.13 per gasoline-equivalent litre (*i.e.* adjusting for the lower energy content of ethanol) and very high well-to-wheels GHG reductions per litre, already provides reductions at a cost of \$20 to \$60 per tonne, by far the lowest-cost biofuels option.

Thus, another key finding of this book is that, at least in the near term, the costs of producing biofuels are much lower in tropical and subtropical

This graph is questionable
(location of the fulls)

Figure 2

Biofuels Cost per Tonne of Greenhouse Gas Reduction

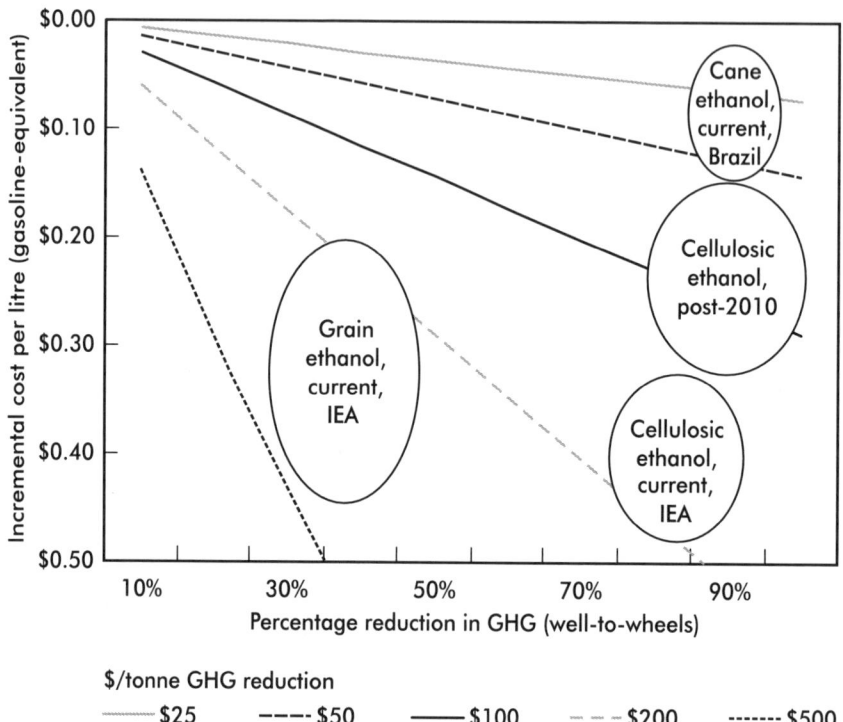

Note: Approximate range of cost per tonne of CO_2-equivalent reduction in well-to-wheels GHG emissions, taking into account ethanol incremental cost per litre and GHG reduction per litre.
Source: IEA estimates – see Chapter 4.

countries – especially developing countries with low land and labour costs – than in developed, temperate countries (*e.g.* most IEA countries). However, there is a mismatch between those countries where biofuels can be produced at lowest cost and those where demand for biofuels is rising most rapidly. If biofuels needs and requirements of IEA countries over the next decade were met in part with a feedstock base expanded beyond their borders, then the costs of biofuels could drop substantially, and their potential for oil displacement (on a global basis) could increase substantially.

Global Potential for Biofuels Production

Chapter 6 assesses land requirements and land availability for producing biofuels. Scenarios developed for the United States and the European Union indicate that near-term targets of up to 6% displacement of petroleum fuels with biofuels appear feasible using conventional biofuels, given available cropland. A 5% displacement of gasoline in the EU requires about 5% of available cropland to produce ethanol, while in the US 8% is required. A 5% displacement of diesel requires 13% of US cropland, 15% in the EU. Land requirements for biodiesel are greater primarily because average yields (litres of final fuel per hectare of cropland) are considerably lower than for ethanol. Land requirements to achieve 5% displacement of *both* gasoline and diesel would require the combined land total, or 21% in the US and 20% in the EU. These estimates could be lower if, for example, vehicles experience an efficiency boost running on low-level biofuels blends and thus require less biofuel per kilometre of travel[1].

At some point, probably above the 5% displacement level of gasoline and diesel fuel, biofuels production using current technologies and crop types may begin to draw substantial amounts of land away from other purposes, such as production of crops for food, animal feed and fibre. This could raise the price of other commodities, but it could also benefit farmers and rural communities. Chapter 4 reviews several recent analyses of the impact of biofuels production on crop prices. The impacts can be significant at even fairly low levels of biofuels production. More work in this area is clearly needed to establish a better understanding of the effects of biofuels production on other markets. The potential for biofuels production in IEA countries is much greater if new types of feedstocks (*e.g.* cellulosic crops, crop residues, and other types of biomass) are also considered, using new conversion technologies.

The potential global production of biofuels for transport is not yet well quantified. Our review of recent studies reveals a wide range of long-term estimates of bioenergy production potential for all purposes – including household energy use, electricity generation and transportation. Even using the more conservative estimates, it appears that a third or more of road

---

1. *As discussed in Chapter 5, there is no consensus in the literature on biofuels impacts on vehicle efficiency. In this book, equal energy efficiency of vehicles running on petroleum fuels and on biofuels-blends is assumed unless otherwise noted.*

transportation fuels worldwide could be displaced by biofuels in the 2050-2100 time frame. However, most studies have focused on technical rather than economic potential, so the cost of displacing petroleum fuel associated with most estimates is very uncertain. Further, use of biomass for transport fuels will compete with other uses, such as for heat and electricity generation, and it is not yet clear what the most cost-effective allocations of biomass are likely to be.

One recent study focuses on the near-term potential for economically competitive cane ethanol production worldwide through 2020. The study estimates that enough low-cost cane-derived ethanol could be produced over this time frame to displace about 10% of gasoline and 3% of diesel fuel worldwide. However, this ethanol would mostly be produced in developing countries, while demand would be mainly in developed countries (where transport fuel consumption is much higher). Thus, in order to achieve such a global displacement, a substantial international trade in ethanol would need to arise. While this is just one study, focusing on one type of feedstock, it suggests that much more attention should be paid to the global picture, and to the potential role of biofuels trade. Currently many IEA countries have import tariffs on liquid biofuels. To date, the World Trade Organization (WTO) has not looked into issues related to opening up international trade of biofuels.

The Importance of Developing Advanced Biomass-to-Biofuels Conversion Technologies

One potential source of increased biofuels supply in all countries is dedicated energy crops, *i.e.* cellulosic energy crops and crop residues (often called "biomass"), as well as other waste products high in cellulose, such as forestry wastes and municipal wastes. A large volume of crops and waste products could be made available in many countries without reducing the production of food crops, because much land that is not suitable for food crop production could be used to produce grasses and trees. Cellulosic feedstocks could be used to produce ethanol with very low "well-to-wheels" greenhouse gases, since they can be converted to ethanol using lignin (*i.e.* the non-cellulose part of the plant) and excess cellulose instead of fossil fuels as the main process fuel. This new approach would nearly eliminate the need for fossil energy inputs into the conversion process. But advanced conversion technologies are needed to efficiently convert cellulose to alcohol and other fuels such as synthetic diesel, natural gas or even hydrogen in a cost-effective manner. Two key areas of research are under way in IEA countries:

■ **Conversion of cellulose to sugars.** A number of countries, and particularly the United States, have ongoing research programmes to improve technologies to convert cellulose to sugars (in order to then be fermented into alcohol). However, to date the targeted cost reductions for cellulosic ethanol have not been realised, and it appears that, although the first large-scale facilities are to be constructed in the next few years, the cost of this ethanol will still be well above the long-term targeted level. It is unclear to what extent this is due to underfunding of research, to simply needing more time for development, or to inherent limitations in technology, though constructing large-scale, semi-commercial facilities will be an important step. Emphasis in the US biofuels research programme has shifted somewhat since 2000. Recent work has focused on developing test facilities that produce a variety of outputs in addition to biofuels, such as co-generated electricity, chemicals, and possibly food and/or fibre products. These "biorefineries" use cellulose (and lignin) as the primary inputs and process fuel the way current refineries use petroleum. Biorefineries are expected to improve overall conversion efficiencies and the variety and value of outputs for a given input. Greater emphasis is also being placed on developing new strains of crops, including genetically modified crops, as well as new conversion enzymes that can provide higher yields and better conversion efficiency.

■ **Conversion of biomass to transport fuels through gasification and thermo-chemical routes.** A different vein of research is being pursued in a number of IEA countries (in part under the framework provided by the IEA's Bioenergy Implementing Agreement). This approach focuses on technologies to, for example, gasify biomass and use the resulting gases to produce a number of different fuels – including methanol, ethanol, dimethyl ether (DME – an LPG-like fuel suitable for diesel engines), and synthetic diesel and gasoline fuels. It is also possible to use gaseous fuel directly in vehicles. Both methane and hydrogen can be produced through biomass gasification, though these fuels would not be compatible with today's vehicles and would need major modifications to existing fuel infrastructure systems. There are also some approaches not involving gasification – for example creating "biocrude" through high-temperature/ pressure and chemical breakdown of biomass into liquids, using hydrothermal upgrading (HTU) or pyrolysis. The suite of different pathways for producing these "BTLs" (biomass-to-liquids) generally can achieve very high conversion efficiencies, but they are currently expensive and technically immature. It is unclear whether the gasification or other approaches under investigation can achieve cost reductions sufficient to be competitive with other transport fuels over the next 10 to 15 years.

Policy-related Conclusions and Recommendations

The following points summarise this book's major policy-related conclusions and recommendations:

- **Biofuels may be easier to commercialise than other alternative fuels, considering performance, infrastructure and other factors.** Biofuels have the potential to leapfrog traditional barriers to entry because they are liquid fuels largely compatible with current vehicles and blendable with current fuels. In fact, low-percentage ethanol blends, such as E10 (10% ethanol by volume), are already dispensed in many service stations worldwide, with almost no incompatibility with materials and equipment. Thus, biofuels could be used in today's vehicles to reduce global petroleum consumption by 10% or more.

- **Biofuels can play a significant role in climate change policy and in measures to reduce greenhouse gas emissions.** Biofuels have become particularly intriguing because of their potential to greatly reduce CO_2 emissions throughout their fuel cycle. Virtually all of the CO_2 emitted by vehicles during combustion of biofuels does not contribute to new emissions, because the CO_2 is already part of the fixed carbon cycle (absorbed by plants during growth and released during combustion). Moreover, some combinations of biofuel feedstock and conversion processes, such as enzymatic hydrolysis of cellulose to produce ethanol, which uses biomass as the process fuel, can reduce well-to-wheels CO_2-equivalent GHG emissions to near zero.

- **Biofuels use in IEA countries and around the world is increasing rapidly, driven largely by government policies.** Given the current high cost of biofuels compared to petroleum fuels, it is unlikely that widespread use of biofuels will occur without strong policy intervention. However, given the existing high gasoline and diesel taxes around Europe and in many other countries, and lower taxes for biofuels in many countries (with direct subsidies in North America), only relatively minor "tweaks" in policy may be needed to spur the market for biofuels to higher levels. For example, in the United States, the existing subsidy (of about $0.14 per litre) is sufficient to encourage substantial production and sales of corn-derived ethanol as a fuel. An adjustment to this subsidy to vary payments according to the net oil displacement or GHG reduction of the production process could provide a strong incentive for changes in production practices and development of new technologies and feedstocks that would lower well-to-wheels GHGs, and perhaps reduce the costs of these fuels, considerably.

- **Biofuels policies in many countries are largely agriculture-driven.** Current policies related to biofuels in many IEA countries, and particularly in the EU, appear to be driven largely by agricultural concerns, perhaps more than by energy concerns. Agricultural policy in many countries is complex and serves multiple policy objectives. Major producer support schemes are in place around the IEA. Although the OECD does not support the use of agricultural subsidies, it is nonetheless likely that support schemes will continue to play an important role in the future, including for crop feedstocks for biofuels. Some studies have shown that the cost of subsidising increased biofuels production will be at least partly offset by resulting reductions in other agricultural subsidies (for example, set-aside land payments might be reduced if these lands are used to produce biofuels). As promoting biofuels rises on political agendas, agricultural policies will need to be more closely reconciled with energy, environmental, trade and overall economic policies and priorities. This area deserves more analysis than it has received so far.

- **A better understanding of how biofuels production affects crop and food markets is needed.** As mentioned above, while the impact of increased biofuels production on farm income is expected to be mainly positive (due to increases in crop sales and possibly crop prices), the net market impact on all groups is less clear. For example, the impact on consumers could be negative if crop (and food) prices rise due to lower availability of non-biofuels crops (although many IEA countries are currently experiencing crop surpluses). Several recent economic studies indicate that increased production of biofuels could lead to price increases not only of crops used for biofuels, but also of other crops – as land is shifted towards greater production of crops for biofuels production. However, the commercialisation of cellulosic-based ethanol could alleviate price pressures while giving farmers new sources of income, since it would open up new land (like low-value grazing lands) to crop production, and also allow greater productivity from existing cropland (*e.g.* through use of crop residues for biofuels production).

■ **The development of international markets for biofuels could increase benefits and lower costs.** Nearly all analysis and policy initiatives to date in IEA countries have focused on domestic production and use of conventional biofuels. However, there are fairly wide ranges of feedstock availability and production costs among countries and regions. In particular, production costs of sugar cane ethanol in Brazil are much lower than grain ethanol in IEA countries. This cost difference is likely to persist as ethanol production facilities are built in other warm, developing countries, such as India. These cost differences create opportunities for biofuels trade that would substantially lower their cost and increase their supply in IEA countries, and would encourage development of a new export industry in developing countries. Further, since both greenhouse gas emissions and oil import dependence are essentially global problems, it makes sense to look at these problems from an international perspective. For example, IEA countries could invest in biofuels production in countries that can produce them more cheaply, if the benefits in terms of oil use and greenhouse gas emissions reductions are superior to what could be achieved domestically. In a carbon-trading framework such as that being developed with the Clean Development Mechanism under the Kyoto Protocol, biofuels production in developing countries could be a promising source of emissions reduction credits.

■ **The global potential for biofuels production and displacement of petroleum appears substantial.** The global potential of biofuels supply is just beginning to be carefully studied, under various assumptions regarding land availability and other factors. Studies reviewed in Chapter 6 indicate that, after satisfying global food requirements, enough land could be available to produce anywhere from a modest fraction to all of projected global demand for transport fuels over the next 50 years. Relatively low-cost sugar-cane-to-ethanol processes might be able to displace on the order of 10% of world gasoline use in the near term (*e.g.* through 2020); if cellulose-to-ethanol processes can meet cost targets, a far higher percentage of petroleum transport fuels could cost-effectively be replaced with biofuels. Ultimately, advanced biomass-to-liquids processes might provide the most efficient (and therefore least land-intensive) approach to producing biofuels, but costs will need to come down substantially for this to occur.

■ **Many questions remain.** Throughout the book, a number of areas have been identified where further research is needed. Some of the most important are: better quantifying biofuels' various benefits and costs; developing energy and agricultural policy that maximises biofuels-related benefits at minimum government (subsidy) and societal cost; gaining a better and more detailed understanding of global biofuels production potential, cost, and environmental impacts; and applying greater levels of support for research, development, and commercialisation of advanced biofuels production technologies.

1 INTRODUCTION

Improving energy security, decreasing vehicle contributions to air pollution and reducing or even eliminating greenhouse gas emissions are primary goals compelling governments to identify and commercialise alternatives to the petroleum fuels currently dominating transportation. Over the past two decades, several candidate fuels have emerged, such as compressed natural gas (CNG), liquefied petroleum gas (LPG) and electricity for electric vehicles. These fuels feature a number of benefits over petroleum, but they also exhibit a number of drawbacks that limit their ability to capture a significant share of the market. For example, they all require costly modifications to vehicles and the development of separate fuel distribution and vehicle refuelling infrastructure. As a result, except in a few places both fuel suppliers and vehicle manufacturers have been reluctant to make the required investments in this uncertain market.

Biofuels have the potential to leapfrog traditional barriers to entry because they are liquid fuels compatible with current vehicles and blendable with current fuels. They share the long-established distribution infrastructure with little modification of equipment. In fact, low-percentage ethanol blends, such as E10 (10% ethanol by volume), are already dispensed in many service stations worldwide, with almost no incompatibility with materials and equipment. Biodiesel is currently blended with conventional diesel fuel in many OECD countries, ranging from 5% in France to 20% in the US, and is used as a neat fuel (100% biodiesel) in some trucks in Germany.

Expanding the use of biofuels would support several major policy objectives:

- *Energy security.* Biofuels can readily displace petroleum fuels and, in many countries, can provide a domestic rather than imported source of transport fuel. Even if imported, ethanol or biodiesel will likely come from regions other than those producing petroleum (*e.g.* Latin America rather than the Middle East), creating a much broader global diversification of supply sources of energy for transport.

- *Environment.* Biofuels are generally more climate-friendly than petroleum fuels, with lower emissions of CO_2 and other greenhouse gases over the complete "well-to-wheels" fuel chain[1]. Either in their 100% "neat" form or more commonly as blends with conventional petroleum fuels, vehicles running on biofuels emit less of some pollutants that exacerbate air quality problems, particularly in urban areas. Reductions in some air pollutants are also achieved by blending biofuels, though some other types of emissions (*e.g.* NO_x) might be increased this way.

- **Fuel quality.** Refiners and car manufacturers have become very interested in the benefits of ethanol in order to boost fuel octane, especially where other potential octane enhancers, such as MTBE, are discouraged or prohibited.

- **Sustainable transportation.** Biofuels are derived from renewable energy sources.

This book provides an assessment of the potential benefits and costs of producing biofuels in IEA countries and in other regions of the world. Many IEA governments have implemented or are seriously considering new policy initiatives that may result in rapid increases in the use of biofuels. The assessment presented here, of recent trends and current and planned policies, indicates that world production of biofuels could easily double over the next few years. Since there is a great deal of interest and policy activity in this area, and knowledge about biofuels is evolving rapidly, the primary objective of this book is to inform IEA member governments[2] and other policy-makers about the characteristics, recent research, developments, and potential benefits and costs of biofuels at this important policy-making time. Another objective is to identify uncertainties and to urge countries to put more resources into studying them, in order to assist in the development of rational policies towards a more sustainable transportation future.

What are Biofuels?

For many, biofuels are still relatively unknown. Either in liquid form such as fuel ethanol or biodiesel, or gaseous form such as biogas or hydrogen, biofuels

---

1. *"Well-to-wheels" refers to the complete chain of fuel production and use, including feedstock production, transport to the refinery, conversion to final fuel, transport to refuelling stations, and final vehicle tailpipe emissions.*

2. *IEA members include the United States, Canada, twenty European countries, Japan, South Korea, Australia and New Zealand.*

are simply transportation fuels derived from biological (*e.g.* agricultural) sources:

- Cereals, grains, sugar crops and other starches can fairly easily be fermented to produce ethanol, which can be used either as a motor fuel in pure ("neat") form or as a blending component in gasoline (as ethanol or after being converted to ethyl-tertiary-butyl-ether, ETBE).

- Cellulosic materials, including grasses, trees, and various waste products from crops, wood processing facilities and municipal solid waste, can also be converted to alcohol. But the process is more complex relative to processing sugars and grains. Techniques are being developed, however, to more effectively convert cellulosic crops and crop wastes to ethanol. Cellulose can also be gasified to produce a variety of gases, such as hydrogen, which can be used directly in some vehicles or can be used to produce synthesis gas which is further converted to various types of liquid fuels, such as dimethyl ether (DME) and even synthetic gasoline and diesel.

- Oil-seed crops (*e.g.* rapeseed, soybean and sunflower) can be converted into methyl esters, a liquid fuel which can be either blended with conventional diesel fuel or burnt as pure biodiesel.

- Organic waste material can be converted into energy forms which can be used as automotive fuel: waste oil (*e.g.* cooking oil) into biodiesel; animal manure and organic household wastes into biogas (*e.g.* methane); and agricultural and forestry waste products into ethanol. Available quantities may be small in many areas, but raw materials are generally low cost or even free. Converting organic waste material to fuel can also diminish waste management problems.

Global Biofuel Production and Consumption

This book focuses primarily on ethanol and biodiesel. Ethanol is by far the most widely used biofuel for transportation worldwide – mainly due to large production volumes in the US and Brazil. Fuel ethanol produced from corn has been used as a transport fuel in the United States since the early 1980s, and now provides over 10 billion litres (2.6 billion gallons) of fuel per year, accounting for just over 2% of the total US consumption of motor gasoline on a volume basis (about 1.4% on an energy basis). The US production of fuel ethanol is over 20 times greater than production in any other IEA country and, as shown in Figure 1.1, is rising rapidly. In Brazil, production of fuel ethanol from sugar cane began in 1975. Production peaked in 1997 at 15 billion litres, but declined to 11 billion in 2000, as a result of shifting policy goals and measures. Production of ethanol is rising again, however, and still exceeds US production. All gasoline sold in Brazil contains between 22% and 26% ethanol by volume.

Figure 1.1

World and Regional Fuel Ethanol Production, 1975-2003 (million litres per year)

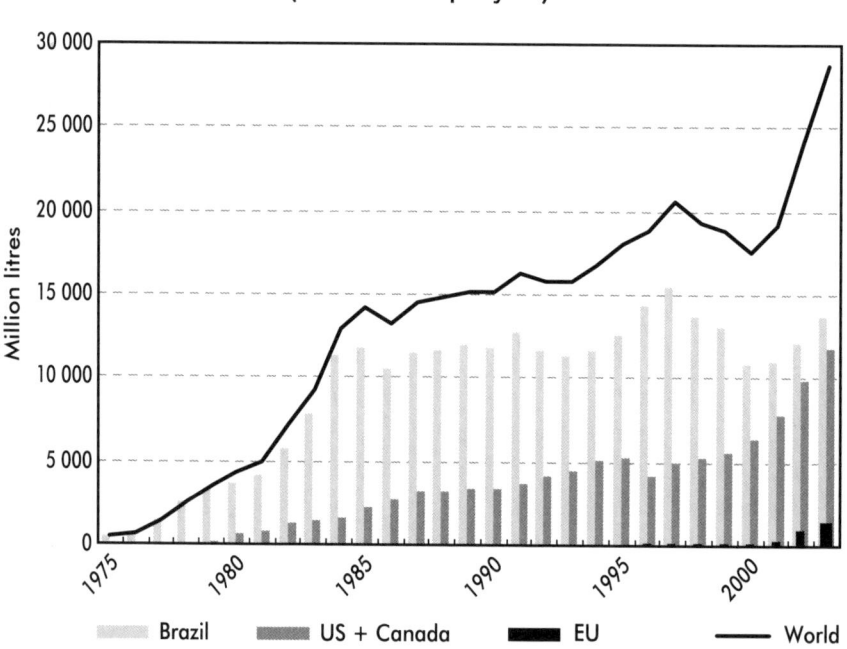

Source: F.O. Lichts (2003). Does not include beverage ethanol production.

As shown in Figure 1.2, biodiesel production is highest in Europe, where more biodiesel is produced than fuel ethanol, but total production of both fuels is fairly small compared to production of ethanol in Brazil and the United States. Only capacity data, not production data, are available for biodiesel, but production is typically a high percentage of capacity. A small amount of European biodiesel is used for non-transportation purposes (*e.g.* for stationary heat and power applications).

Figure 1.2

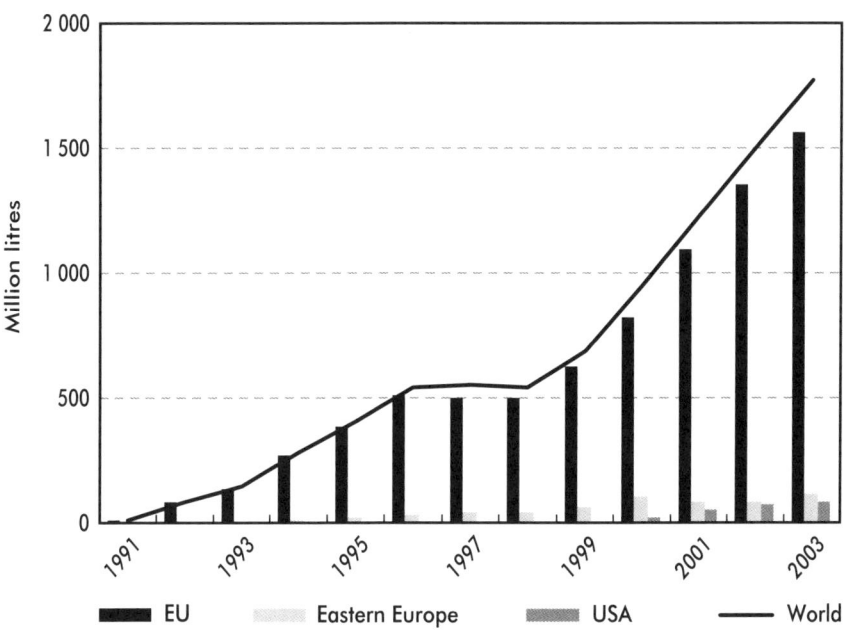

World and Regional Biodiesel Capacity, 1991-2003
(million litres per year)

EU   Eastern Europe   USA   —— World

Note: EU biodiesel production was about two-thirds of capacity in 2003. Source: F.O. Lichts (2003).

Production estimates for 2002 by country and by fuel are shown in Table 1.1. The table also shows typical uses and feedstock in IEA countries. In Europe, the principal biodiesel-producing countries are France, Germany, and Italy. The fuel is used mainly as a diesel blend, typically 5% or 20%. However, in Germany, biodiesel is commonly sold in its 100% "neat" form, and dispensed in some 700 filling stations. Some European vehicle manufacturers have approved the use of 100% biodiesel in certain engines (*e.g.* VW, BMW), while others have been concerned about vehicle/fuel compatibility issues and potential NO_x emission increases from pure biodiesel, and have limited their warranties to cover only lower-level blends (Nylund, 2000).

As discussed in Chapter 7, there have been many recent efforts to expand the use of biofuels in both IEA and non-IEA countries. In early 2003, the European Commission (EC) issued a directive promoting the use of biofuels and other renewable fuels for transport. This directive created two "indicative" targets for

Table 1.1

World Ethanol Production and Biodiesel Capacity, 2002
(million litres)

	Ethanol			Biodiesel	
	Fuel ethanol production	Typical use	Feedstock	Biodiesel production capacity[a]	Typical use
United States	8 151	E10 blends; some E85	corn	70	blends <25%
Canada	358	E10	wheat		
IEA North America	**8 509**			**70**	
Austria				32	blends <25%
Belgium				36	
Denmark				3	
France	117	converted to ETBE	70% beet; 30% wheat	386	mainly 5% blends
Germany				625	100% biodiesel; some blends
Italy				239	blends <25%
Spain	144	converted to ETBE	barley	9	
Sweden	44	blends E10, E85 and E95	wheat	17	blends <25%
UK				6	
Other EU	85				
EU	**390**			**1 353**	
Poland	48	low-level blends		80	
IEA Europe	**438**			**1 433**	
IEA Pacific (just Australia)[b]	40	blends E10 to E20	wheat, cane		
Latin America (just Brazil)	12 620[c]	blends up to E26; dedicated (E95)			
Asia (just China)	289	low-level blends			
World	**21 841**			**1 503**	

[a] Feedstock in the US is from soy, in Europe, rapeseed and sunflower. Production of biodiesel in 2003 is roughly 65% of capacity.
[b] Ethanol blends in Australia restricted to maximum E10 beginning in 2003.
[c] Ethanol tracking in Latin America may include some beverage alcohol.
Sources: F.O. Lichts (2003). Some minor production (e.g. India, Africa) not reported.

EU member states: 2% biofuels penetration by December 2005 and 5.75% by December 2010. The targets are not mandatory, but governments are required to develop plans to meet them. In the US and Canada, legislation is under consideration that could lead to several-fold increases in biofuels (especially ethanol) production over the next few years. Australia has recently implemented blending targets and Japan has made clear its interest in biofuels blending, even if biofuels must be imported. Several non-IEA countries, such as India and Thailand, have recently adopted pro-biofuels policies. In Latin America, major new production capacity is being developed, in part with an eye towards providing exports to an emerging international market in biofuels.

The following chapters cover various aspects of biofuels, focusing primarily on ethanol and biodiesel, but also considering advanced fuels and conversion technologies. Chapter 2 provides a technical review of biofuels production processes. Chapter 3 assesses the potential energy and greenhouse gas impacts of using biofuels. Chapter 4 covers biofuels production and distribution costs, and, drawing on Chapter 3, provides a discussion of biofuels costs per tonne of CO_2-equivalent greenhouse gas reductions under various assumptions. Chapter 5 covers issues related to vehicle/fuel compatibility and infrastructure. Chapter 6 reviews recent assessments of land use requirements for liquid biofuels production, and the consequent production potential given current technology and the available land resource base in North America, the EU and worldwide. Estimates of the global potential for oil displacement and greenhouse gas reductions are provided. Chapter 7 reviews recent policy activity in various countries around the world, and provides a projection of biofuels production over the next 20 years, given the policies and targets that have been put in place. Chapter 8 provides a discussion of policy-related issues and recommendations for additional research on this topic.

2 FEEDSTOCK AND PROCESS TECHNOLOGIES

This chapter reviews the various processes available for producing transportation-grade biofuels from sugar, grain, cellulosic and oil-seed crop feedstock. While a wide variety of approaches and feedstock exist for producing ethanol, there is a much narrower range for biodiesel. Emerging techniques to gasify virtually any type of biomass and turn this gas into almost any type of liquid fuel create a variety of new possibilities. We look first at biodiesel production from oil-seed crops, followed by ethanol production from several different feedstock types and processes, and finally look at emerging techniques for gasifying biomass and producing various finished fuels.

Biodiesel Production

The term "biodiesel" generally refers to methyl esters (sometimes called "fatty acid methyl ester", or FAME) made by transesterification, a chemical process that reacts a feedstock oil or fat with methanol and a potassium hydroxide catalyst[1]. The feedstock can be vegetable oil, such as that derived from oil-seed crops (*e.g.* soy, sunflower, rapeseed, etc.[2]), used frying oil (*e.g.* yellow grease from restaurants) or animal fat (beef tallow, poultry fat, pork lard). In addition to biodiesel, the production process typically yields as co-products crushed bean "cake", an animal feed, and glycerine. Glycerine is a valuable chemical used for making many types of cosmetics, medicines and foods, and its co-production improves the economics of making biodiesel. However, markets for its use are limited and under high-volume production scenarios, it could end up being used largely as an additional process fuel in making biodiesel, a relatively low-value application. Compared with some of the technologies being developed to produce ethanol and other biofuels, the biodiesel production process involves well-established technologies that are

---

1. *"Biodiesel" also includes synthetic diesel fuel made from biomass through gasification or other approach. This is discussed later in the chapter. Throughout this book, "FAME" (from "fatty acid methyl ester") is used to refer specifically to biodiesel from transesterification of oils and fats.*
2. *Soy is often called soya, particularly in Europe, while rapeseed oil is referred to as "canola" in Canada.*

The Biodiesel Production Process

Biodiesel from fatty acid methyl esters (FAME) can be produced by a variety of esterification technologies, though most processes follow a similar basic approach. First the oil is filtered and pre-processed to remove water and contaminants. If free fatty acids are present, they can be removed or transformed into biodiesel using pre-treatment technologies. The pre-treated oils and fats are then mixed with an alcohol (usually methanol) and a catalyst (usually sodium or potassium hydroxide). The oil molecules (triglycerides) are broken apart and reformed into esters and glycerol, which are then separated from each other and purified. The resulting esters are biodiesel.

---

not likely to change significantly in the future. Biodiesel can be used in compression ignition diesel systems, either in its 100% "neat" form or more commonly as a 5%, 10% or 20% blend with petroleum diesel.

Ethanol Production

Ethanol can be produced from any biological feedstock that contains appreciable amounts of sugar or materials that can be converted into sugar such as starch or cellulose. Sugar beets and sugar cane are obvious examples of feedstock that contain sugar. Corn, wheat and other cereals contain starch (in their kernels) that can relatively easily be converted into sugar. Similarly, trees and grasses are largely made up of cellulose and hemicellulose, which can also be converted to sugar, though with more difficulty than conversion of starch.

Ethanol is generally produced from the fermentation of sugar by enzymes produced from yeast. Traditional fermentation processes rely on yeasts that convert six-carbon sugars (mainly glucose) to ethanol. Because starch is much easier than cellulose to convert to glucose, nearly all ethanol in northern countries is made from widely-available grains. The organisms and enzymes for starch conversion and glucose fermentation on a commercial scale are readily available. Cellulose is usually converted to five- and six-carbon sugars, which requires special organisms for complete fermentation. The key steps in the feedstock-to-ethanol conversion process, by feedstock type, are shown in Figure 2.1 and discussed in the following sections.

Figure 2.1

Ethanol Production Steps by Feedstock and Conversion Technique

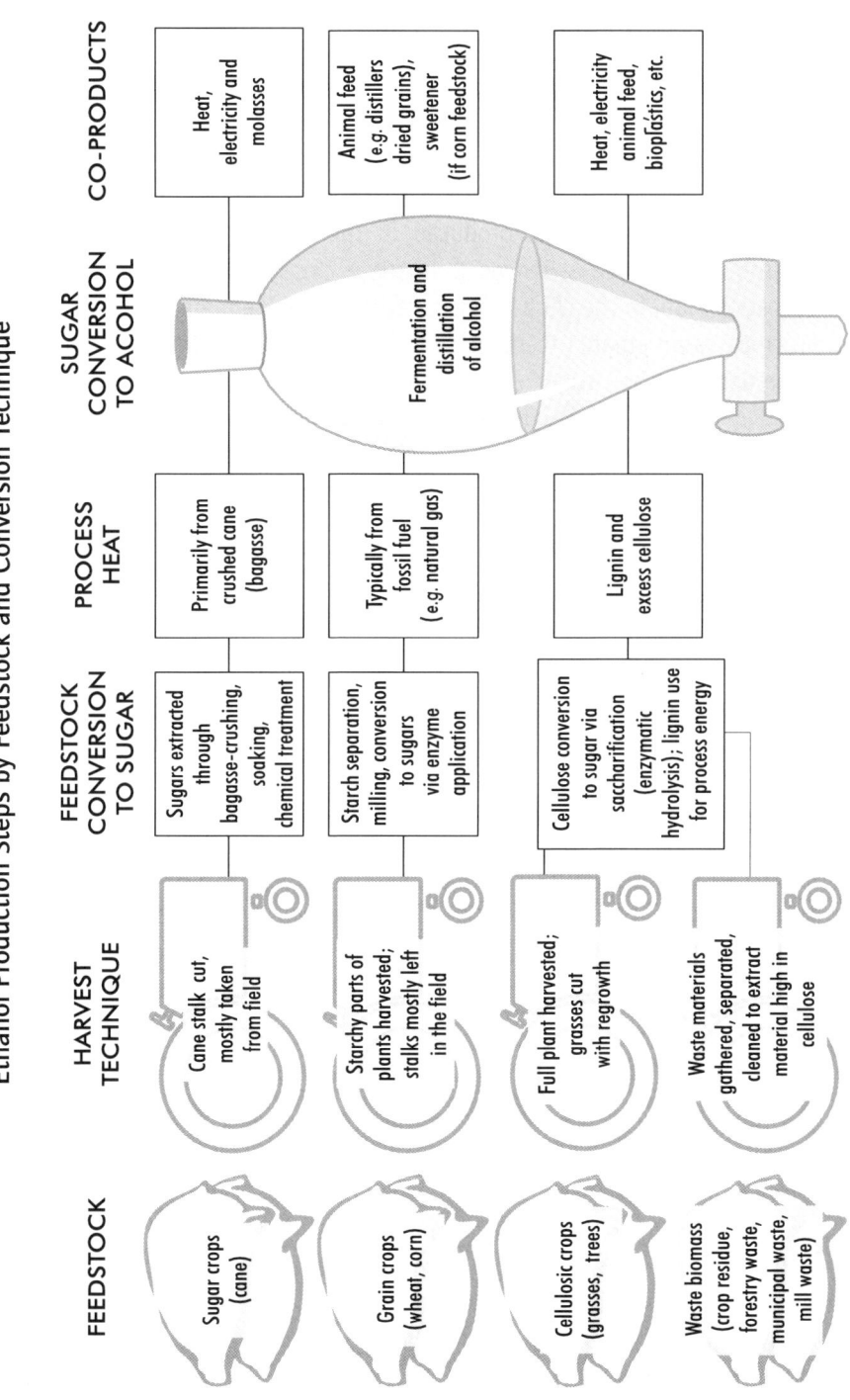

Sugar-to-Ethanol Production

The least complicated way to produce ethanol is to use biomass that contains six-carbon sugars that can be fermented directly to ethanol. Sugar cane and sugar beets contain substantial amounts of sugar, and some countries in the EU (*e.g.* France) rely on sugar beet to produce ethanol. Until the 1930s, industrial-grade ethanol was produced in the United States through the fermentation of molasses derived from sugar crops. However, the relatively high cost of sugar in the US has since made sugar cane more expensive than grain crops as an ethanol feedstock. In Brazil and in most tropical countries that produce alcohol, sugar cane is the most common feedstock used to produce ethanol. As discussed in Chapter 4, costs of ethanol production from sugar cane in warm countries are among the lowest for any biofuels.

The Sugar-to-Ethanol Production Process

In producing ethanol from sugar crops, the crops must first be processed to remove the sugar (such as through crushing, soaking and chemical treatment). The sugar is then fermented to alcohol using yeasts and other microbes. A final step distils (purifies) the ethanol to the desired concentration and usually removes all water to produce "anhydrous ethanol" that can be blended with gasoline. In the sugar cane process, the crushed stalk of the plant, the "bagasse", consisting of cellulose and lignin, can be used for process energy in the manufacture of ethanol. As discussed in Chapter 3, this is one reason why the fossil energy requirements and greenhouse gas emissions of cane-to-ethanol processes are relatively low.

Grain-to-Ethanol Production

In IEA countries, most fuel ethanol is produced from the starch component of grain crops (primarily corn and wheat in the US and wheat and barley in Europe – though sugar beets are also used in Europe). In conventional grain-to-ethanol processes, only the starchy part of the crop plant is used. When corn is used as a feedstock, only the corn kernels are used; for wheat, it is the whole wheat kernel. These starchy products represent a fairly small percentage of the total plant mass, leaving considerable fibrous remains (*e.g.* the seed husks and stalks of these plants). Current research on cellulosic ethanol production (discussed below) is focused on utilising these waste cellulosic materials to create fermentable sugars – ultimately leading to more efficient production of ethanol than from using just the sugars and starches directly available.

The Grain-to-Ethanol Production Process

The grain-to-ethanol production process starts by separating, cleaning and milling (grinding up) the starchy feedstock. Milling can be "wet" or "dry", depending on whether the grain is soaked and broken down further either before the starch is converted to sugar (wet) or during the conversion process (dry). In both cases, the starch is converted to sugar, typically using a high-temperature enzyme process. From this point on, the process is similar to that for sugar crops, where sugars are fermented to alcohol using yeasts and other microbes. A final step distils (purifies) the ethanol to the desired concentration and removes water. The grain-to-ethanol process also yields several co-products, such as protein-rich animal feed (e.g. distillers dry grain soluble, or DDGS) and in some cases sweetener, although this varies depending on the specific feedstock and process used.

Cellulosic Biomass-to-Ethanol Production

Most plant matter is not sugar or starch, but cellulose, hemicellulose and lignin. The green part of a plant is composed nearly entirely of these three substances[3]. Cellulose and hemicellulose can be converted into alcohol by first converting them into sugar (lignin cannot). The process, however, is more complicated than converting starch into sugars and then to alcohol.

Today, there is virtually no commercial production of ethanol from cellulosic biomass, but there is substantial research going on in this area in IEA countries, particularly the US and Canada. There are several potentially important benefits from developing a viable and commercial cellulosic ethanol process:

■ Access to a much wider array of potential feedstock (including waste cellulosic materials and dedicated cellulosic crops such as grasses and trees), opening the door to much greater ethanol production levels.

---

3. *Throughout this book, the term "cellulosic" is used to denote materials high in cellulose and hemicellulose.*

■ Greater avoidance of conflicts with land use for food and feed production.

■ A much greater displacement of fossil energy per litre of fuel, due to nearly completely biomass-powered systems.

■ Much lower net well-to-wheels greenhouse gas emissions than with grain-to-ethanol processes powered primarily by fossil energy.

A large variety of feedstock is available for producing ethanol from cellulosic biomass. The materials being considered are agricultural wastes (including those resulting from conventional ethanol production), forest residue, municipal solid wastes (MSW), wastes from pulp/paper processes and energy crops. Agricultural wastes available for ethanol conversion include crop residues such as wheat straw, corn stover (leaves, stalks and cobs), rice straw and bagasse (sugar cane waste). Forestry wastes include underutilised wood and logging residues; rough, rotten and salvable dead wood; and excess saplings and small trees. MSW contains some cellulosic materials, such as paper and cardboard. Energy crops, developed and grown specifically for fuel, include fast-growing trees, shrubs, and grasses such as hybrid poplars, willows and switchgrass. The cellulosic components of these materials can range anywhere from 30% to 70%. The remainder is lignin, which cannot be converted to sugar, but can be used as a process fuel in converting cellulose to alcohol, or can be converted to liquid fuel through gasification and gas-to-liquids conversion (see following section).

In terms of production potentials, forest and agricultural residue sources, such as corn stover, represent a tremendous resource base for biomass ethanol production and, in the long term, could support substantial growth of the ethanol industry. For example, as shown in Chapter 6, in the US stover could provide more than ten times the current ethanol production derived from grains. In Brazil, sugar cane stalks ("bagasse") are used to provide process energy for ethanol conversion, after the sugar is removed, but this cellulosic material is not yet converted into ethanol itself. Further, much of the sugar cane crop is usually left in the field, and commonly burned. Thus, even though Brazilian ethanol already shows excellent greenhouse gas reduction and cost characteristics (as described in Chapters 3 and 4), more complete use of cellulosic components could improve Brazilian processes further.

Dedicated energy crops such as switchgrass, hybrid willow and hybrid poplar provide an important feedstock option. Switchgrass is typically grown on a ten-year crop rotation basis, and harvest can begin in year 1 in some locations and year 2 in others. Harvests involve mowing and collecting the grass, requiring no annual replanting or ploughing. Trees such as willows and hybrid poplar require more time, up to 10 years to reach harvest age, depending upon the growing region. The use of grasses and woody crops opens up a much greater range of land areas that can be used for growing energy crops than for sugar and grain crops, and requires lower energy (fertiliser) input. It may also provide improved wildlife habitats compared to row-crop farming (Murray, 2002).

To convert cellulose to ethanol, two key steps must occur. First, the cellulose and hemicellulose portions of the biomass must be broken down into sugars through a process called saccharification. The yielded sugars, however, are a complex mixture of five- and six-carbon sugars that provide a greater challenge for complete fermentation into ethanol. Second, these sugars must be fermented to make ethanol, as they are in grain-to-ethanol processes. The first step is a major challenge, and a variety of thermal, chemical and biological processes are being developed to carry out this saccharification step in an efficient and low-cost manner (see box).

One important difference between cellulosic and conventional (grain and sugar crop) ethanol production is the choice of fuel to drive the conversion process. This choice has important implications for the associated net energy balances and for net greenhouse gas emissions (discussed in Chapter 3). In current grain-to-ethanol production processes in North America and Europe, virtually all process energy is provided by fossil inputs, such as natural gas used to power boilers and fermentation systems. For cellulose-to-ethanol conversion, nearly all process energy is provided by biomass, in particular the unused cellulosic and lignin parts of the plant being processed. Given current grain harvesting practices in North America and Europe, only relatively small amounts of non-starch components are easily available for process fuel. In short, it has been easier and less expensive to continue relying on fossil energy inputs to drive the conversion process, even though this emits far more greenhouse gases than conversion relying on bioenergy as the process fuel.

A number of research organisations and companies are exploring combinations of thermal, chemical and biological saccharification processes to develop the most efficient and economical route for the commercial production of cellulosic ethanol. These programmes have substantial government support, particularly in the United States and Canada. None of the approaches, however, has as yet been demonstrated on a large-scale,

The Cellulosic Biomass-to-Ethanol Production Process

The first step in converting biomass to ethanol is pre-treatment, involving cleaning and breakdown of materials. A combination of physical and chemical (e.g. acid hydrolysis) processes is typically applied, which allows separation of the biomass into its cellulose, hemicellulose and lignin components. Some hemicellulose can be converted to sugars in this step, and the lignin removed.

Next, the remaining cellulose is hydrolysed into sugars, the major saccharification step. Common methods are dilute and concentrated acid hydrolysis, which are expensive and appear to be reaching their limits in terms of yields. Therefore, considerable research is being invested in the development of biological enzymes that can break down cellulose and hemicellulose. The first application of enzymes to wood hydrolysis in an ethanol process was to simply replace the cellulose acid hydrolysis step with a cellulose enzyme hydrolysis step. This is called separate hydrolysis and fermentation (SHF). An important process modification made for the enzymatic hydrolysis of biomass was the introduction of simultaneous saccharification and fermentation (SSF), which has recently been improved to include the co-fermentation of multiple sugar substrates. In the SSF process, cellulose, enzymes and fermenting microbes are combined, reducing the required number of vessels and improving efficiency. As sugars are produced, the fermentative organisms convert them to ethanol (Sreenath et al., 2001).

Finally, researchers are now looking at the possibility of producing all required enzymes within the reactor vessel, thus using the same "microbial community" to produce both the enzymes that help break down cellulose to sugars and to ferment the sugars to ethanol. This "consolidated bioprocessing" (CBP) is seen by many as the logical end point in the evolution of biomass conversion technology, with excellent potential for improved efficiency and cost reduction (Hamelinck et al., 2003).

commercially viable level. Millions of research dollars are going into improving enzymatic hydrolysis processes, mostly targeting improved process efficiencies and yields. The largest of these programmes is in the US, where reducing enzyme costs by a factor of ten and improving the effectiveness of biomass pre-treatment are major goals. If these goals can be achieved, vast amounts of low-cost, low greenhouse gas emitting, high feedstock potential ethanol could become available to fuel markets worldwide.

Research on Cellulosic Ethanol in the United States and Canada

With the advent of new tools in the field of biotechnology, researchers have succeeded in producing several new strains of yeast and bacteria that exhibit varying degrees of ability to convert the full spectrum of available sugars to ethanol. However, the development of cellulosic ethanol technology has been hampered by technical problems associated with the separation of cellulose from lignin and the conversion of cellulose to sugars. Therefore, concentrating research on developing more efficient separation, extraction and conversion techniques is crucial to increase ethanol production.

The US Department of Energy operates a research programme that in FY 2003 had a budget of over $100 million for biomass-related activities (DOE, 2002a). A significant share of this was devoted to research programmes for use of cellulosic feedstock to produce liquid fuels, as well as in "biorefinery" applications to produce multiple products including transport fuels, electric power, chemicals and even materials such as plastics. Current aspects of the US Department of Energy's efforts include:

- **Biomass feedstock "infrastructure".** Characterisation of the physical and mechanical properties of crop residues and analysis of alternative processes for increasing the bulk density of biomass for transport; development of novel harvesting equipment designs, storage and logistics[4].

- **Feedstock conversion research.** A key research area is "bioprocessing", which involves combining different types of enzymes, and genetically engineering new enzymes, that work together to release both hemicellulosic sugars and cellulosic sugars in an optimal fashion. The National Renewable Energy Laboratory (NREL) operates a small (one tonne per day) process development unit, where bioethanol developers can test proposed processes under industrial conditions without having to build their own pilot plants.

---

4. *The US Department of Energy no longer funds research directly into the development of energy crops and instead relies on the US Department of Agriculture to sponsor this type of work.*

■ **Biorefinery projects.** The US Dept. of Energy biomass programme recently awarded about US$ 75 million in six major cost-sharing agreements for integrated biomass research and development, to investigate various approaches to developing biorefineries and to construct test facilities (DOE, 2002b)[5]. The programme's present goal is to produce cellulosic ethanol in biorefineries on a commercial scale by 2010. However, in this time frame, the net cost of producing cellulosic ethanol is not expected to fall to that of gasoline. Ethanol produced from cellulosic biomass is expected to become more competitive as larger and more advanced commercial scale biorefineries are built between 2010 and 2020.

In Canada, ethanol research and development has been carried primarily through the Renewable Energy Technologies Program (RETP), which is managed by the CANMET Energy Technology Center (CETC). RETP supports efforts by Canadian industry to develop and commercialise advanced renewable energy technologies that can serve as cost-effective and environmentally responsible alternatives to conventional energy generation. The focus of the biofuels efforts within the RETP has been the conversion of plentiful and inexpensive cellulosic biomass to ethanol and value-added chemicals. The programme supports pilot-scale projects such as Queen's University's extractive fermentation process and Tembec Inc.'s hemicellulose fermentation efforts. The intent is to demonstrate technology developed under the programme and to promote its transfer to the private sector.

Canadian government support has also been given to Iogen Corporation of Ottawa to further develop an integrated process for the production of fuel ethanol from cellulosic feedstock such as wood waste. Iogen has constructed a demonstration facility in Ottawa that can process up to four thousand tonnes of wheat straw per year, producing up to one million litres of ethanol. Iogen has indicated plans to construct a full-size conversion facility (of the "biorefinery" type) within a few years, through a strategic partnership with Shell, at an as-yet undetermined location. Such a facility might be able to process several hundred times the amount of feedstock of the current test facility, but would need sufficient feedstock supplies within the area – typically a 100-150 kilometre radius (EESI, 2003).

---

5. *Some current multiple-product ethanol conversion plants already exist, and have been called biorefineries, but these all use grain starch, not cellulose, as the primary input. Future cellulose and lignin-based biorefineries may be capable of producing many more types of products, with better overall economics.*

IEA Research in Bioenergy

Since its creation in 1974, the IEA has promoted international co-operation in energy technology R&D and deployment, providing a legal framework for joint R&D networks (called "Implementing Agreements"). These have helped achieve faster technological progress and innovation at lower cost, helped reduce R&D risks and avoid the duplication of effort.

For biofuels research, the most relevant network is the Implementing Agreement on Bioenergy (www.ieabioenergy.com). It co-ordinates the work of national programmes across the wide range of bioenergy technologies and has 19 countries participating. The network covers all aspects of bioenergy, and their ongoing task on liquid biofuels involves working jointly with governments and industry to identify and eliminate non-technical environmental and institutional barriers that impede the use of liquid fuels from biomass in the transport sector. In addition, it aims to identify remaining technical barriers to liquid biofuel technologies and recommend strategies for overcoming these barriers, to consolidate these efforts and formulate a deployment strategy (www.forestry.ubc.ca/task39).

Another relevant network is the IEA Implementing Agreement on Advanced Motor Fuels. This network runs an "Automotive Fuels Information Service" and has a range of projects under way (http://www.vtt.fi/virtual/amf/). For information about IEA networks in related areas and on joining the networks, see http://www.iea.org/about/cert.htm.

Biomass Gasification and Related Pathways

Another approach, or suite of approaches, to converting biomass into liquid or gaseous fuels is direct gasification, followed by conversion of the gas to final fuel. Ethanol can be produced this way, but other fuels can be produced more easily and potentially at lower cost, though none of the approaches is currently inexpensive. Possible target fuels include methanol, synthetic diesel and gasoline (the latter two produced using the "Fischer-Tropsch" process to build the carbon-chain molecules), dimethyl ether (DME – a potential alternative fuel for diesel engines with good combustion properties and low emissions), and gaseous fuels such as methane (CH_4) and hydrogen. DME and the gaseous fuels are not compatible with today's gasoline or diesel vehicles and would need both new types of vehicles (such as compressed natural gas or hydrogen fuel cell vehicles) and new refuelling infrastructure. In all cases, biomass can be converted into final fuels using biomass-derived heat and electricity to drive the conversion process, resulting in very low well-to-wheels greenhouse-gas fuels.

There are a variety of processes available both for the biomass gasification step and for converting this gas into a final fuel. IEA's Implementing Agreement on Bioenergy (www.ieabioenergy.com) is helping to co-ordinate research in this area among IEA members and some non-member countries.

The simplest gasification process converts biomass into methane (commonly called "biogas"). In many countries, biogas "digestion" facilities are commonly used by households, farms and municipalities, with the methane providing energy for cooking and other heat applications. These all operate on the principle of creating the right conditions for bacterial breakdown of biomass and conversion to methane, typically using anaerobic digestion. Two prevalent types of digesters are the Chinese "fixed dome" digester and the Indian "floating cover" digester, which differ primarily in the way gas is collected and routed out of the digester (ITDG, 2000).

Although anaerobic digestion is the best-known and best-developed technology for biochemical conversion of biomass into biogas, new technologies are being developed in IEA countries such as the Netherlands, Germany and Japan (Novem, 2003). These new systems can be specially designed to produce a variety of different gases and end products. They typically use heat and/or chemicals to break down biomass into gas, with little or no microbial action involved. Most approaches fall into either the heat or chemical-dominated categories. The choice of which process to use is influenced by the fact that lignin cannot easily be converted into a gas through biochemical conversion (just as it cannot be converted into alcohol). Lignin can, however, be gasified through a heat process. The lignin components of plants can range from near 0% to 35%. For those plants at the lower end of this range, the chemical conversion approach is better suited. For plants that have more lignin, the heat-dominated approach is more effective.

Depending on the process, a number of different gases may be released, including methane, carbon monoxide and dioxide, nitrogen and hydrogen. Much research is now focused on maximising the hydrogen yield from such processes, for example in order to provide hydrogen for fuel cell applications (see box). Once the gasification of biomass is complete, the resulting gases can be used in a variety of ways to produce liquid fuels, including:

- **Fischer-Tropsch (F-T) fuels.** The Fischer-Tropsch process converts "syngas" (mainly carbon monoxide and hydrogen) into diesel fuel and naphtha (basic gasoline) by building polymer chains out of these basic building blocks. Typically a variety of co-products (various chemicals) are also produced. Finding markets for these co-products is essential to the economics of the F-T process, which is quite expensive if only the gasoline and diesel products are considered (Novem, 2003).

- **Methanol.** Syngas can also be converted into methanol through dehydration or other techniques, and in fact methanol is an intermediate product of the F-T process (and is therefore cheaper to produce than F-T gasoline and diesel). Methanol is somewhat out of favour as a transportation fuel due to its relatively low energy content and high toxicity, but might be a preferred fuel if fuel cell vehicles are developed with on-board reforming of hydrogen (since methanol is an excellent hydrogen carrier and relatively easily reformed to remove the hydrogen).

- **Dimethyl ether.** DME also can be produced from syngas, in a manner similar to methanol. It is a promising fuel for diesel engines, due to its good combustion and emissions properties. However, like LPG, it requires special fuel handling and storage equipment and some modifications of diesel engines, and is still at an experimental phase. Its use has only been tested in a few diesel vehicles. If diesel vehicles were designed and produced to run on DME, they would become inherently very low pollutant-emitting vehicles; with DME produced from biomass, they would also become very low GHG vehicles.

Regardless of the final fuel or the process, gasification methods are still being developed and are currently expensive. As discussed in Chapter 4, it appears that all techniques for biomass gasification and conversion to liquid fuels are as or more expensive than enzymatic hydrolysis of cellulose to sugar, followed by fermentation. With both types of approaches, costs will need to come down

Hydrogen from Biomass Production Processes

Production of hydrogen for vehicles could become very important if hydrogen fuel cell vehicles become commercialised in the future. Biomass could provide a very low GHG source of hydrogen, even serving as a conduit for returning CO_2 from the atmosphere into the earth, if biomass gasification to hydrogen were combined with carbon sequestration (Read, 2003).

In traditional biochemical conversion (digestion) processes, wet feedstock such as manure is digested for 2-4 weeks to produce primarily CH_4 and CO_2. To produce hydrogen, the CH_4 has to be converted using a thermochemical process, such as steam reforming. By manipulation of process conditions, methane formation can be suppressed and hydrogen can be directly produced along with organic acids. These acids can then be converted into methane and post-processed to yield additional hydrogen, increasing the overall efficiency of the process. Overall, this approach is well developed, though innovations to increase efficiency and lower costs are still needed in order to bring the cost of hydrogen production with this method closer to that of hydrogen production from other sources (such as direct reforming of natural gas).

Thermochemical conversion processes are at an earlier stage of development, and a variety of approaches are being tested, nearly all of which include a gasification step. Gasification typically involves using heat to break down the biomass and produce a "synthesis gas", often composed of several compounds from which the H_2 must afterwards be extracted. Gasification can be conducted using a variety of low, medium or high-temperature methods. These methods differ in several respects, including required pre-treatment (pyrolysis or torrefaction may be needed to partially break down the biomass and convert it into a form that is fed more easily to the gasifier) and post-gasification treatment (it may be steam reformed or partially oxidised, along with a "water-gas-shift" reaction to extract H_2 from the synthesis gas).

substantially – by at least half – in order for these fuels to compete with petroleum fuels at current world oil prices.

Besides gasification, there are several other innovative approaches to produce transportation fuels from biomass. One of these is diesel production through hydrothermal upgrading (HTU). In contrast to gasification technologies, the HTU process consists of dissolving cellulosic materials in water under high pressure, but relatively low temperature. Subsequent reactions convert the feedstock into a "biocrude" liquid. It is subsequently upgraded to various hydrocarbon liquids, especially diesel fuel, in a hydrothermal upgrading unit (Schindler and Weindorf, 2000). Another approach uses "fast pyrolysis", whereby biomass is quickly heated to high temperatures in the absence of air, and then cooled down, forming a liquid ("bio-oil") plus various solids and vapours. This oil can be further refined into products such as diesel fuel. The approach is also used to convert solid biomass residues such as "bagasse" (sugar cane residue) into a fuel that is easier to burn for process heat during production of ethanol (DESC, 2001). Like the gasification approaches, these processes are still under development and in need of substantial cost reduction in order to become economic.

Achieving Higher Yields: the Role of Genetic Engineering

One way to increase the benefits and lower the costs of producing biofuels is to raise crop yields. High crop yields per acre and per energy input (like fertiliser) reduce cost, increase potential biofuels supply, and significantly improve the well-to-wheels greenhouse gas characteristics of the final fuel. Although traditional methods like selective breeding continue to play the main role in improving crop yields, biotechnology offers an important new approach, particularly in the mid to long term.

Genetically engineered (GE) crops, also known as genetically modified organisms (GMOs), have genes from other species inserted or substituted in their genomes. Transgenic transfers give a plant different characteristics very quickly. Traditional plant breeding techniques use a range of natural variability in plant species to increase productivity and hybrid vigour, but rates of improvement can be slow. Gene mapping can be used both for traditional breeding (to speed up the selection process) and for developing GE crops (Peelle, 2000).

Although there is still considerable uncertainty, genetically modified crop genomes are expected to lead to major increases in yields, reductions in fertiliser requirements and improvements in pest-resistance. The greatest barriers are the social concerns regarding their safety in the food chain.

Dedicated energy crops such as switchgrass may in fact face fewer obstacles than food crops, since they are not consumed by humans.

Research into plant "genomics" (the study of genes) now amounts to several hundred million dollars per year, worldwide, with the vast majority of funding and activity occurring in the private sector (NRC, 1999). Genomics technology can lead to the development of new, high-yield, pest-resistant varieties of plants and can enable major modifications to the production characteristics and feedstock quality that would be very difficult to achieve through traditional breeding. Once the necessary genes are available in the gene pools of bioenergy crops, genetic engineering could also be used to produce new co-products.

Genetic research into food crops has already resulted in new high-yielding varieties of corn, wheat and sugar cane. Yet the most important advances for biofuels may occur in dedicated energy crops like cellulose-rich trees and grasses. Genetic research into dedicated energy crops is at a much earlier stage, however. Current research is focused on mapping gene sequences and identifying key "markers", *i.e.* locations where modifying genetic code could provide significant benefits (Dinus, 2000). Research for dedicated energy crops is more reliant on government funding than research for food/feed crops, since there are fewer near-term markets for energy crops. Major advances in genetic mapping, gene function studies and field trials of newly created materials will not likely occur before 2010. The US National Research Council has called for substantially increased government funding (NRC, 1999), in order to push forward the date when new, improved varieties will be ready for production systems.

Most of the current bioenergy crop research is focused on switchgrass and poplar, because they have particular advantages that should facilitate the application of biotechnology. Switchgrass is closely related to rice, corn and sugar cane, organisms that are also being intensively studied. The entire rice genome is being sequenced by an international public consortium. The genetic similarity between the switchgrass genome and these food-crop genomes should facilitate the gene-level analysis of switchgrass. Tree species like poplar and willow have no close relatives under genomic study. However, these species have a number of traits that could facilitate genomic studies. They have small genomes, simplifying gene identification and mapping. In addition, several pedigrees already exist for poplars as a result of breeding programmes and ongoing genomic studies, and poplars can be readily transformed by methods of asexual gene transfer; thus they have given rise to many more transgenic plants than any other woody species.

Genetic engineering can result in a higher percentage of cellulose/hemicellulose (and thus a lower lignin content) in dedicated energy crops and a greater uptake of carbon in root systems. The creation of new types of co-products is also possible. Co-products under commercial development via the genetic engineering of plants include vaccines and other high-value pharmaceuticals, industrial and specialty enzymes, and new fragrances, oils and plastics. Bioenergy crops could also be engineered to produce large quantities of cellulytic enzymes, which could be used directly for feedstock processing and thus reduce the cost of cellulase required for feedstock conversion to ethanol and co-products. For other feedstocks, such as corn stover and switchgrass, genetic engineering methods will be available from major biotechnology companies as a result of their work on maize and rice. More possibilities will be revealed as catalogues of genes in lignocellulosic tissues are uncovered by genomics studies.

The analysis in this book, particularly in Chapter 6, is not based on the extensive use of genetic engineering to increase plant yields. However, a steady improvement in yields is assumed, on the order of 1% per year, consistent with recent trends for food crops in IEA countries. Much faster improvement may be possible in the future, if GE crops are developed and deployed – or it may turn out that genetic engineering becomes necessary just to maintain historical yield improvement rates.

3 OIL DISPLACEMENT AND GREENHOUSE GAS REDUCTION IMPACTS

Recent events around the world have once again put energy security, and in particular oil import dependence, at the top of energy agendas in IEA countries. The emergence of global climate change as a critical energy and environmental policy issue has also heightened awareness that combustion of greenhouse gas-emitting fossil fuels imposes risks for the planet. Biofuels may provide a partial solution to each of these problems, by displacing oil use in transport and by reducing greenhouse gas (GHG) emissions per litre of fuel consumed.

Estimating the net impacts of using biofuels on oil use and GHG emissions is a complex issue, however, and requires an understanding of fuel compositions, fuel production methods, combustion processes and related technologies throughout the full "fuel cycle", from biomass feedstock production to final fuel consumption. In order to address these complexities and to determine the approaches that are most likely to maximise oil and other fossil energy displacement and to reduce overall GHG emissions, there is a growing body of literature on estimating the full fuel cycle or "well-to-wheels" greenhouse gas emissions.

While such analysis is being conducted for a variety of potential alternative fuels for transport, biofuels have become particularly intriguing to researchers because of their potential to greatly reduce CO_2 emissions throughout the fuel cycle. Virtually all CO_2 emitted during vehicle combustion of biofuel does not contribute to new emissions of carbon dioxide, because the emissions are already part of the fixed carbon cycle (absorbed by plants during growth and released during combustion).

A key question for biofuels is how much CO_2 and other GHG emissions are released during all phases of fuel production. In some cases, emissions may be as high or higher than the net GHG emissions from gasoline vehicles over the gasoline fuel cycle. On the other hand, some combinations of biofuel feedstock and conversion processes, such as enzymatic hydrolysis of cellulose to produce ethanol and using biomass as the process fuel, can reduce well-to-wheels CO_2-equivalent GHG emissions to near zero.

Research on the net GHG reduction impacts of biofuels is progressing, but is far from conclusive, and the variation in results from different studies provides insights for understanding the impacts of different approaches to producing and using biofuels. Most of the studies evaluate grains (for ethanol) and oil crops (for biodiesel) in North America and the EU, but a few have looked at sugar crops, including sugar cane in Brazil. The following sections cover each of these fuels, feedstock and regions.

Ethanol from Grains

Most studies over the past ten years indicate that, compared to gasoline, the use of ethanol derived from grains, using currently commercial processes, brings a 20% to 40% reduction in well-to-wheels CO_2-equivalent GHG emissions (Table 3.1). The one exception is a study by Pimentel (2001), and some of the reasons for its different estimates are outlined below.

One of the most important assumptions driving these estimates is the overall fuel production process efficiency – how much process fuel is required to grow crops, transport them to distilleries, produce ethanol and deliver it to refuelling stations. Studies that estimate better process efficiencies (represented by a lower number in Table 3.1) tend to have greater GHG reduction estimates. The feedstock-to-ethanol conversion plant efficiency is an important factor in determining the overall process energy use, as it determines how much feedstock must be grown, moved and processed to produce a given volume of ethanol.

Beyond process efficiency, there are several other potentially important factors, though analysis of some of them is difficult because several of the studies do not provide detailed findings or assumptions. For example, it is useful to know the assumptions regarding "co-product credits", *i.e.* the amount of energy and GHG emissions that co-products of ethanol production process, such as animal feed and co-generated electricity, help displace by reducing the production of competing items. All the studies in Table 3.1 except Pimentel's assume that the production of various co-products reduces the net GHG impact of corn ethanol by 5% to 15%.

The type of process energy used – particularly for feedstock conversion into ethanol – is also important. Most of the North American studies make similar

Table 3.1

Energy and GHG Impacts of Ethanol:
Estimates from Corn- and Wheat-to-Ethanol Studies

	Feedstock	Ethanol production efficiency (litres/tonne feedstock)	Fuel process energy efficiency (energy in/out)	Well-to-wheels GHG emissions: compared to base (gasoline) vehicle (per km travelled)	
				Fraction of base vehicle	Percent reduction
GM/ANL, 2001	corn-*a*	372.8	0.50	n/a	n/a
GM/ANL, 2001	corn-*b*	417.6	0.55	n/a	n/a
Pimentel, 2001/91	corn	384.8	1.65	1.30	–30%[c]
Levelton, 2000	corn	470.0	0.67	0.62	38%
Wang, 2001a	corn-dry mill	387.7	0.54	0.68	32%
Wang, 2001a	corn-wet mill	372.8	0.57	0.75	25%
Levy, 1993	corn-*a*	367.1	0.85	0.67	33%
Levy, 1993	corn-*b*	366.4	0.95	0.70	30%
Marland, 1991	corn	372.8	0.78	0.79	21%
Levington, 2000	wheat	348.9	0.90	0.71	29%
ETSU, 1996	wheat	346.5	0.98	0.53	47%
European Commission, 1994	wheat	385.4	1.03	0.81	19%
Levy, 1993	wheat-*a*	349.0	0.81	0.68	32%
Levy, 1993	wheat-*b*	348.8	0.81	0.65	35%

Note: Where a range of estimates is reported by a paper, "*a*" and "*b*" are shown in the feedstock column to reflect this.
[c] Negative greenhouse gas reduction estimate connotes an increase. n/a: not available.
Sources: Except for Levelton, 2000, Wang 2001a and GM/ANL 2001, data presented here for these studies are taken from the comparison conducted by CONCAWE, 2002.

assumptions regarding the average share of process energy derived from oil, gas and coal, both directly and for generation of electricity used in ethanol production. Relatively little oil is used, though considerable quantities of gas, and even some coal use, is generally assumed. Oil is mainly used to run farm equipment and to transport feedstock and final fuel to their destinations. It rarely amounts to more than 20% of the energy contained in the final ethanol fuel, so that production and use of one litre of grain ethanol typically displaces about 0.8 or more litres of gasoline, on an energy-equivalent basis.

Finally, estimates of vehicle fuel economy (*e.g.* fuel consumption in megajoules per kilometre) are important for the GHG comparison when analysing the net impacts per vehicle kilometre driven (rather than just per unit of fuel produced). Different assumptions regarding the *relative* efficiency of gasoline and ethanol (or fuel-blended) vehicles can have a significant impact (up to a 10% variation) on results. Studies typically range from assuming the same efficiency (kilometres per unit energy of fuel), such as the GM/ANL (2001) study, to assuming up to a 10% energy efficiency gain from dedicated (or E95) vehicles, such as Wang (2001a).

The European studies of ethanol production from wheat estimate a similar range of GHG reduction potential as the North American studies for corn. The ETSU study (1996), however, estimates a 47% reduction potential from wheat. Process energy efficiency and conversion efficiency estimates for wheat tend to be lower (higher energy use) than for corn, so the reason for the similar or better GHG reduction estimates is not completely clear. It appears to be a function of the type of process energy used in the United Kingdom (primarily natural gas).

Another important consideration is the types of greenhouse gases considered, and assumptions regarding their impact on the climate. Nearly all studies include carbon dioxide (CO_2), nitrous oxide (N_2O), and methane (CH_4). However, many do not look at ozone (O_3), which also affects climate directly, but is not emitted as such from fuel cycles. Rather, its concentration is influenced by other gases that are emitted from fuel cycles, *e.g.* nitrogen oxides (NO_x), carbon monoxide (CO) and non-methane organic compounds (NMOCs). Since 1990, NO_x, CO, and NMOCs have been identified as "indirect" GHGs because of their effects on ozone (Delucchi, 1993). The Intergovernmental Panel on Climate Change (IPCC) provides "global warming potential" (GWP) factors for these compounds. More recently, aerosols have been identified as direct and indirect GHGs, and work is proceeding to identify which kinds or components of aerosols affect climate most (Delucchi, 2003). Most recently, hydrogen (through its effect on ozone) and particulate matter (including black carbon) have been identified as potential GHGs. The number of GHGs undoubtedly will continue to grow as researchers identify more direct and indirect greenhouse gases.

Assumptions regarding nitrogen (as N_2O) in the crop production system also can have an important impact on estimates. The natural absorption and release patterns of nitrogen by plants and soils, and the use of nitrogen fertilisers, have received considerable treatment in recent studies. The GM *et al.* study (2002) for Europe shows in some detail how different practices regarding fertiliser use, and different assumptions and estimates regarding their rate of N_2O release (as well as the uncertain nature of N_2O as a greenhouse gas) can dramatically change the overall estimate of CO_2-equivalent reductions[1]. There is still much uncertainty regarding the full effects of nitrogen after it leaves the site of crop production and enters the atmosphere.

Assumptions for land use and land use change are also important. For example, if crops are planted on land that was or would otherwise become a forest, then there is a significant emission of GHGs associated with the loss of carbon sequestration. Even the manner in which a bioenergy crop fits into crop rotation cycles can have significant impacts, such as on the net release or absorption of N_2O.

Very recently, Delucchi (2004) has explored the impacts of land use change on biofuels' GHG emissions. His preliminary findings are that in cases where land is brought into production of certain crops for biofuels, depending on the previous use of the land there may be an important one-time impact on greenhouse gas emissions. This is due to release of greenhouse gases from changes in soil conditions, changes from root systems of different plants, etc. For example, in the case of corn-to-ethanol, if a significant fraction of the land was previously covered with prairie grass, Delucchi estimates that the GHG emissions related to land use change might be as large as emissions from the fuel production (corn-to-ethanol) stage. In this case, he estimates that corn-to-ethanol would have slightly higher total (well-to-wheels) CO_2-equivalent emissions than gasoline (averaged over a number of years). The effect may be even larger in the case of biodiesel, because much more land is required to generate a unit of fuel than in the case of corn. Delucchi points out that these findings are preliminary and dependent on uncertain assumptions, and he has not yet looked at a full range of crop types[2]. The effect of land use change is clearly an area where more attention and research are needed.

---

1. *This study's estimates are based on ethanol from sugar beets and cellulose, which is discussed in the following sections.*
2. *Dr. Delucchi requested that his earlier well-to-wheels estimates for biofuels not be reported in the tables in this chapter, since he is in the process of revising them to include new data on land use change.*

Table 3.2

Net Energy Balance from Corn-to-Ethanol Production: A Comparison of Studies

	Energy content[a]	Corn crop yield	Conversion efficiency	Ethanol produced per acre of corn crop	Nitrogen fertiliser energy	Ethanol conversion process	Total energy used to produce litre	Co-products energy credits	Net energy value	Energy in/out (with co-product credit)
	Btu/litre	bushels/acre	litres/bushel	litres/acre (000)	MBtu/litre	MBtu/litre	MBtu/litre	MBtu/litre	MBtu/litre	ratio
Ho, 1989	287.7	90	n/a	n/a	n/a	215.7	340.7	39.7	-15.1	1.04
Marland and Turhollow, 1990	317.8	119	9.46	1.13	50.3	189.6	279.8	30.8	68.7	0.80
Pimentel, 1991	287.7	110	9.46	1.04	70.3	278.9	495.9	81.4	-126.9	1.34
Keeney and DeLuca, 1992	287.7	119	9.69	1.15	63.7	183.5	345.2	30.6	-31.9	1.08
Shapouri et al., 1995	317.8	122	9.58	1.17	34.0	201.7	313.5	57.0	61.3	0.84
Lorenz and Morris, 1995	317.8	120	9.65	1.16	42.0	204.2	306.9	104.4	115.8	0.73
Wang et al., 1999	287.7	125	9.65	1.21	32.8	154.6	259.1	56.6	85.2	0.75
Agri-Food Canada, 1999	287.7	116	10.18	1.18	n/a	190.8	259.1	53.2	112.9	0.76
Pimentel, 2001	287.7	127	9.46	1.20	51.6	284.3	496.1	81.4	-127.0	1.34
Shapouri, Duffield and Wang, 2002	317.8	125	10.07	1.26	27.0	196.0	292.3	54.4	79.9	0.79

[a] The numbers in this column (either 287.7 or 317.8) reflect whether high heating value or low heating value data for ethanol were used in the analysis. n/a: not available. MBtu: million British thermal units.
Source: Table from Shapouri et al. (2002); some of the individual studies were consulted and are listed in the references.

Although major uncertainties remain, the available estimates of the GHG emissions reduction potential from starchy (grain) crops suggest that they can provide significant reductions. The trend appears to be that the more recent the estimate, the greater the estimated reduction, as crop yields and process efficiencies continue to improve. Other big improvements could come from increasing the use of bioenergy in the conversion process. For example, it is possible to use the residues (*e.g.* "straw") from grain crops as process fuel (much like is done in Brazil with sugar cane bagasse, discussed below). This could improve the fossil energy balance and GHG emissions picture considerably. However, this is not currently done at any large-scale grain-to-ethanol plants.

The Net Energy Balance of Corn-to-Ethanol Processes

There has been considerable discussion recently regarding the net energy gain from producing ethanol from grains. Some research has suggested that it may take more fossil energy to produce a litre of ethanol (i.e. to grow, harvest and transport the grain and convert it to ethanol) than the energy contained in that litre. This would suggest that conversion losses wipe out the benefit of the renewable energy (i.e. sunlight) used to grow the crops. The non-solar energy used in the different stages of the process is primarily natural gas and coal. Only about one-sixth of the fossil energy used to produce grain ethanol in the US is estimated to be petroleum.

A recent report published by the US Department of Agriculture (Shapouri et al., 2002) provides a review of net energy studies of corn-to-ethanol processes over the last ten years. The key assumptions and results of these studies are shown in Table 3.2. The net energy value of ethanol production (energy in the ethanol minus the energy used to produce it) has been found to be both positive and negative by different studies, although most of the more recent estimates show a positive balance. The key factors and assumptions that have varied most across studies are:

- *Corn yield per hectare.*

- *Ethanol conversion efficiency and energy requirements.*

- *Energy embedded in the fertiliser used to grow corn.*

- *Assumptions regarding use of irrigation.*   *(continued)*

> ■ *The value, or "energy credit", given for co-products produced along with ethanol (mainly animal feed).*
>
> *The most pessimistic of the recent studies, Pimentel's (2001), uses a number of older estimates from one of the author's previous studies that do not reflect improvements in aspects such as crop yields and conversion efficiencies during the 1990s. On the other hand, Pimentel has included certain factors absent in other studies, such as the energy embedded in farm equipment and the cement used in ethanol plant construction. These factors, however, account for only a small share of the differences in estimates.*
>
> *Estimates from more recent studies show a fairly narrow range, with one energy unit of ethanol requiring between 0.6 and 0.8 fossil energy units to produce it (taking into account co-product credits). Most of this fossil energy is not petroleum-based. Shapouri et al. (2002) estimate that only about 17% of input energy is from petroleum fuels, with the vast majority from natural gas and coal (including electricity derived from these fuels). Using this estimate, 0.12 to 0.15 energy units of petroleum-based fuels are required to produce one energy unit of ethanol. Put another way, one gasoline-equivalent litre of ethanol displaces 0.85 to 0.88 litres of petroleum on a net energy basis.*

Ethanol from Sugar Beets

Three European studies estimate the GHG reduction potential of producing ethanol from sugar beets (Table 3.3). The studies indicate that this feedstock and conversion process can provide up to a 56% reduction in well-to-wheels GHG emissions compared to gasoline. The same factors that are important in comparing studies for ethanol production from grains apply to studies on production from sugar beets. The wide range in variation of both feedstock production efficiencies and conversion process efficiencies suggests that more work is needed in this area.

Ethanol from Sugar Cane in Brazil

Few studies are available that assess the net energy balance and greenhouse gas emissions from sugar cane ethanol. Those studies that have been done

Table 3.3

Estimates from Studies of Ethanol from Sugar Beets

	Feedstock	Ethanol production efficiency (litres/tonne feedstock)	Fuel process energy efficiency (energy in/out)	Well-to-wheels GHG emissions, compared to base gasoline vehicle (per km travelled)	
				Fraction of base vehicle	Percent reduction
GM *et al.*, 2002	beet	n/a	0.65	0.60	41%
EC, 1994	beet	54.1	0.64	0.50	50%
Levy, 1993	beet-*a*	101.3	0.84	0.65	35%
Levy, 1993	beet-*b*	101.3	0.56	0.44	56%

For Levy estimates, *a* and *b* are high and low process efficiency estimates. n/a: not available.
Sources: CONCAWE (2002), except GM *et al.* (2002).

focus on Brazil, where the process of converting sugar cane to ethanol has improved considerably over the past 20 to 30 years and is now relatively efficient. Sugar cane/ethanol plants in Brazil generally make excellent use of biomass as process energy. Fossil fuel inputs are low compared to grain-to-ethanol processes in the US and Europe.

Macedo *et al.* (2003) and Macedo (2001) are the only two recent, available assessments of the net energy and emissions characteristics of cane ethanol in Brazil (though studies go back as far as Silva *et al.*, 1978). Macedo *et al.* (2003) updates Macedo (2001), mainly by updating much of the data from the 1995-2000 time frame to 2002 for average and best practice plants. The net energy results of comparing average and best are shown in Table 3.4. The balance takes into account energy used during crop production (including fertiliser production), transport, conversion to ethanol and energy used in the construction of all equipment, including conversion plants. Only fossil energy, not renewable energy, is shown in the table. The net energy balance (energy out divided by fossil energy input) is shown to be about 8 on average and 10 in best cases. This means that for each unit of ethanol produced, only about 0.1 units of fossil energy are required, far better than the 0.6-0.8 required to produce a unit of ethanol from grain in the US or Europe. There are two key reasons for this:

■ The sun's rays are intense in Brazil and soil productivity is high, thus sugar cane crop yields are quite high with relatively low fertiliser inputs.

■ Nearly all conversion plant process energy is provided by "bagasse" (the remains of the crushed cane after the sugar has been extracted). Thus, the electricity requirement from fossil fuels in Table 3.4 is zero, while there is excess bagasse energy produced. In fact, many recent plants are designed to co-generate electricity and they are net exporters of energy, resulting in net fossil energy requirements near zero or possibly below zero because the exported electricity is greater than the fossil energy requirements of the process.

The authors point out that the average values for 1995 are considerably higher than 1985 values, including significant improvements in cane yields per hectare of land and conversion yields to ethanol. The 2002 conversion efficiency averaged about 90 litres ethanol per tonne of cane, compared to 85 litres in 1995, and 73 litres in 1985. With a 2002 average harvest yield of

Table 3.4

Energy Balance of Sugar Cane to Ethanol in Brazil, 2002

	Energy requirement (MJ / tonne of processed cane)	
	Average	**Best values**
Sugar cane production	202	192
Agricultural operations	38	38
Cane transportation	43	36
Fertilisers	66	63
Lime, herbicides, etc.	19	19
Seeds	6	6
Equipment	29	29
Ethanol production	49	40
Electricity	0	0
Chemicals and Lubricants	6	6
Buildings	12	9
Equipment	31	24
Total energy input	251	232
Energy output	2 089	2 367
Ethanol	1 921	2 051
Bagasse surplus	169	316
Net energy balance (out/in)	8.3	10.2

Source: Macedo *et al.* (2003).

68.7 tonnes of cane per hectare, this translates into about 6 200 litres per hectare per year. Best values are 10% to 20% better than the average. Given recent trends, the best values in 2002 will likely become the average values in five or ten years. Recent regulations ban the practice of burning dry residual biomass left in the field, so it will be harvested green and will likely be added to bagasse used for energy production. This could further improve the energy balance (Moreira, 2002).

Given the very high rate of energy output per unit of fossil energy input, it is not surprising that well-to-wheels CO_2 emissions are very low. Macedo *et al.* estimate them to be about 92%. Ethanol well-to-wheels CO_2 is estimated to be, on average, about 0.20 kg per litre of fuel used, versus 2.82 kg for gasoline. This takes into account the emissions of CO_2 as well as two other greenhouse gases, methane and N_2O (both mainly released from farming, from the use of fertilisers and from N_2 fixed in the soil by sugar cane then released to the atmosphere).

Ethanol from Cellulosic Feedstock

Several North American studies have focused on the potential for cellulosic feedstock, such as poplar trees and switchgrass, to produce ethanol. As described in Chapter 2, cellulosic materials can be converted to ethanol using enzymatic hydrolysis and related processes that are under intensive research around the IEA. The use of cellulosic feedstock in producing ethanol has a "double value" in that the left over (mainly lignin) parts of the plant can be used as process fuel to fire boiler fermentation systems. This makes for both a relatively energy-efficient production process and a more renewable approach since fossil energy use for feedstock conversion can be kept to a minimum.

For the well-to-wheels estimate of GHG emissions from cellulosic biomass, assumptions regarding the end-use efficiency of vehicles and the amount of fertiliser used to grow the crops become quite important. Variations in these assumptions cause much of the disparity in different estimates of net GHG impacts. The assumption about co-products, including electricity produced by co-generation facilities, is also very important. If the co-generated electricity is used to displace coal-fired power on the grid, this can boost the GHG reduction from the cellulosic process considerably. The net GHG reduction can even be greater than 100%, if the CO_2 absorbed during the growing of the feedstock is greater than the CO_2-equivalent emissions released during the entire well-to-wheels process (taking into account the CO_2 avoided by, for example, displacing high-CO_2 electricity generation).

Typical estimates for net GHG emissions reductions from production and use of cellulosic ethanol are in the range of 70% to 90% compared to conventional gasoline (Table 3.5). The estimates are mainly from engineering studies. Few large-scale production facilities yet exist to obtain more empirically derived estimates or to determine if the assumed efficiencies apply to actual plants. Improvements in cellulosic conversion process efficiency have come more slowly than has been projected over the last decade. But it is nonetheless likely that 70% or better reductions in GHG emissions can be achieved.

Table 3.5

Estimates from Studies of Ethanol from Cellulosic Feedstock

	Feedstock	Ethanol production efficiency (litres/tonne feedstock)	Fuel process energy efficiency (energy in/out)[a]	Well-to-wheels GHG emissions: compared to base (gasoline) vehicle (per km travelled)	
				Fraction of base vehicle	Percent reduction
GM *et al.*, 2002	wood (poplar plantation)	n/a	1.20	0.49	51%
GM/ANL, 2001	wood-*a*	288	1.30	n/a	n/a
GM/ANL, 2001	wood-*b*	371	1.90	n/a	n/a
Wang, 2001a	wood	288	1.52	−0.07	107%
GM/ANL, 2001	grass-*a*	303	1.00	0.29	71%
GM/ANL, 2001	grass-*b*	390	1.60	0.34	66%
Wang, 2001a	grass	303	1.37	0.27	73%
Levelton, 2000b	grass	310	1.28	0.29	71%
GM *et al.*, 2002	crop residue (straw)	N/a	n/a	0.18	82%
Levelton, 2000b	corn residue (stover)	345	1.10	0.39	61%
Levelton, 2000b	hay	305	1.32	0.32	68%
Levelton, 2000b	wheat straw	330	1.12	0.43	57%

Note: Where a range of estimates is reported by a paper, "*a*" and "*b*" are shown in the feedstock column to reflect this. n/a: not available.
[a] Process energy includes both biomass and non-biomass energy sources.
Sources: GM *et al.* (2002), GM/ANL *et al.* (2001), Wang (2001a), and Levelton (2000b).

Biodiesel from Fatty Acid Methyl Esters

Table 3.6 presents findings from studies on the net energy savings, oil savings and well-to-wheels GHG emission impacts from using biodiesel from fatty acid methyl esters (FAME) rather than conventional diesel fuel (typically for truck applications). The European studies generally focus on rapeseed methyl ester (RME), *i.e.* biodiesel from oil-seed rape, while the North American studies look at both rape (called "canola" in Canada) and soy-based biodiesel.

Table 3.6

Estimates from Studies of Biodiesel from Oil-seed Crops

	Feedstock	Ethanol production efficiency (litres/tonne feedstock)	Fuel process energy efficiency (energy in/out)	Well-to-wheels GHG emissions, compared to base diesel vehicle (per km travelled)	
				Fraction of base vehicle	Percent reduction
GM *et al.*, 2002	rape	n/a	0.33	0.51	49%
Levington, 2000	rape	1.51	0.4	0.42	58%
Levelton, 1999	canola (rape)	n/a	n/a	0.49	51%
Altener, 1996	rape-*a*	1.13	0.55	0.44	56%
Altener, 1996	rape-*b*	1.32	0.41	0.34	66%
ETSU, 1996	rape	1.18	0.82	0.44	56%
Levy, 1993	rape-*a*	1.18	0.57	0.56	44%
Levy, 1993	rape-*b*	1.37	0.52	0.52	48%
Levelton, 1999	soy	n/a	n/a	0.37	63%

Note: Where a range of estimates is reported by a paper, "*a*" and "*b*" are shown in the feedstock column to reflect this.
n/a: not available.
Source: All studies from CONCAWE (2002), except GM *et al.* (2002), and Levelton (1999), cited directly.

The estimates for net GHG emissions reductions from rapeseed-derived biodiesel range from about 40% to 60% compared to conventional diesel fuel in light-duty compression-ignition engines. Similar to the findings for ethanol, the range in estimates for biodiesel is explained partly by differences in conversion and energy efficiency assumptions and partly by disparities in assumptions regarding co-product credits.

Most of the studies focus on biodiesel blends of 10% or 20%. The vehicle efficiency of engines running on petroleum diesel fuel or on biodiesel (including various blends) is generally very similar. Thus, the results can be converted to indicate the GHG reductions from the biodiesel itself, not the blend. For example, if biodiesel is estimated to provide a 50% reduction in well-to-wheels GHG emissions, then a 20% blend (B-20) would provide about one-fifth of this, or 10%, per vehicle-kilometre driven.

The recent *GM et al.* (2002) study assesses a number of different cases for rapeseed methyl ester in Europe that vary widely in terms of their GHG emissions, depending on assumptions regarding crop rotation, fertiliser use and the use of the glycerine by-product. Only one GM case is shown in Table 3.6. Co-produced glycerine can be used either to displace other glycerine production or to provide an additional fuel in the biodiesel production process. Compared to using glycerine as a process fuel, the use of glycerine to displace other glycerine production boosts the well-to-wheels estimates for the reduction of net GHG emissions by an additional 30% compared to diesel. Glycerine markets in most countries, however, are likely to be saturated if biodiesel production grows to the point where it accounts for several percentage points of transportation fuel use. The studies assume that additional glycerine beyond this point would then be used as a process fuel.

Other Advanced Biofuels Processes

As discussed in Chapter 2, there are a variety of other advanced technology processes under development to turn biomass into gaseous and liquid fuels that could be used in light and/or heavy-duty vehicles. Studies are emerging that have looked at the potential well-to-wheels impacts of the various technologies. A detailed treatment and comparison of the different processes is beyond the scope of this book, but a basic sense of the energy and CO_2 impacts of these types of processes can be gained from Table 3.7, which presents some of the results of a detailed assessment carried out by the Dutch Energy Agency (Novem), the Dutch Transport Agency, and Arthur D. Little Consulting.

The Novem/ADL study estimated well-to-wheels energy efficiency and CO_2 emissions that might be typical in the 2010-2015 time frame for the processes in Table 3.7. The authors found that nearly all of the pathways provide very

Table 3.7

Estimates of Energy Use and Greenhouse Gas Emissions from Advanced Biofuels from the Novem/ADL Study (1999)

Fuel	Feedstock / location	Process	Well-to-tank		Well-to-wheels	
			Process energy efficiency (energy in/out)	Percent efficiency	CO_2-equivalent GHG emissions g/km	GHG% reduction v. gasoline/diesel
Diesel	petroleum	refining	1.10	91%	198	
Biodiesel	rapeseed (local)	oil to FAME (transesterification)	1.60	62%	123	38%
Biodiesel	soybeans (local)	oil to FAME (transesterification)	1.43	70%	94	53%
Diesel	biomass – eucalyptus (Baltic)	HTU biocrude	1.47	68%	79	60%
Diesel	biomass – eucalyptus (Baltic)	gasification / F-T	2.35	43%	-16	108%
Diesel	biomass – eucalyptus (Baltic)	pyrolysis	3.31	30%	72	64%
DME	biomass – eucalyptus (Baltic)	gasification / DME conversion	1.78	56%	22	89%
Gasoline	petroleum	refining	1.20	83%	231	
Gasoline	biomass – eucalyptus (Baltic)	gasification / F-T	2.71	37%	-10	104%
Ethanol	biomass – poplar (Baltic)	enzymatic hydrolysis (CBP)	1.94	51%	-28	112%
Ethanol	biomass – poplar (Brazil)	enzymatic hydrolysis (CBP)	1.94	51%	-28	112%
Ethanol	biomass – poplar (local with feedstock from Brazil)	enzymatic hydrolysis (CBP)	1.94	51%	-3	101%
Ethanol	corn (local)	fermentation	2.25	45%	65	72%
Hydrogen	biomass – eucalyptus (Baltic)	gasification	2.41	42%	11	95%
CNG	biomass – eucalyptus (local)	gasification	1.69	59%	39	83%

Note: For a discussion of each of the processes listed in this table, see Chapter 2. CBP: combined bioprocessing. F-T: Fischer Tropsch process.
Source: Novem/ADL (1999).

high reductions in well-to-wheels GHG emissions compared to diesel or gasoline vehicles, over 100% in several cases. This is mainly because in every process, biomass provides both the feedstock and much of the process energy for its own conversion (as is the case for cellulosic ethanol). The greatest reductions were found with cellulose-to-ethanol through enzymatic hydrolysis, using the consolidated bioprocessing (CBP) approach[3], but biomass gasification and conversion to final fuels such as diesel and DME provide similar reductions.

The distance that final fuels – and even raw feedstock – are transported has only a minor impact on the well-to-wheels CO_2 emissions of a particular process. For example, ethanol from poplar trees has a similar net CO_2 reduction whether the fuel is produced and used "locally" (*e.g.* in the Netherlands) or whether the ethanol is transported from far away (*e.g.* from Brazil). This indicates that the net energy requirements of long-distance ocean transport of fuels is quite small per litre of fuel shipped. Even shipping raw materials from Brazil to the Netherlands only reduces the net CO_2 emissions by about 10% compared to gasoline.

---

3. *See chapter 2 for a discussion of this process.*

4 BIOFUEL COSTS AND MARKET IMPACTS

Despite continuing improvements in biofuel production efficiencies and yields, the relatively high cost of biofuels in OECD countries remains a critical barrier to commercial development. For "conventional" biofuels, *i.e.* biodiesel from oil-seed crops and ethanol from grain and sugar crops, the technologies involved are fairly mature. While incremental cost reductions can be expected, no major breakthroughs are anticipated that could bring costs down dramatically. Costs will likely continue to decline gradually in the future through technical improvements and optimisations, and as the scale of new conversion plants increases. For these fuels, the cost of feedstock (crops) is a major component of overall costs. This is compounded by the volatility of crop prices. In particular, the cost of producing oil-seed-derived biodiesel is dominated by the cost of the oil and by competition from high-value uses like cooking. Various agricultural subsidy programmes in IEA countries (and the EU) also have significant impacts on crop prices – though the net effects of these programmes are difficult to determine and no attempt is made here to "unravel" them. The size and scale of the conversion facility can also have a substantial impact on costs. The generally larger US conversion plants produce biofuels, particularly ethanol, at lower cost than plants in Europe.

Production costs for ethanol are much lower in countries with a warm climate, with Brazil probably the lowest-cost producer in the world. Production costs in Brazil, using sugar cane as the feedstock, have recently been recorded at less than half the costs in Europe. As discussed in Chapter 6, production of sugar cane ethanol in developing countries could provide a low-cost source for substantial displacement of oil worldwide over the next 20 years. No other type of biofuel shows such near-term potential.

For biofuels produced in IEA countries, the greatest potential cost reductions lie in continued development of advanced technologies to convert biomass (cellulosic) feedstock to ethanol and, eventually, to hydrogen and to other liquid fuels like synthetic diesel. A number of recent studies suggest that the cost of producing cellulosic ethanol could fall below the cost of producing grain ethanol in the 2010-2020 time frame, and may already be cheaper (if large-scale conversion facilities were built) on a cost-per-tonne greenhouse gas reduction basis.

Biofuels Production Costs

This section reviews recent estimates of production costs of biofuels, broken down by region and feedstock type to the extent possible. The focus is on production in North America, the EU, and Brazil, since few data are available for other regions.

Costs of Producing Ethanol from Grain and Sugar Beet in the US and the EU

Currently there is no global market for ethanol, as there is for conventional petroleum fuels. This fact, along with the wide range of crop types, agricultural practices, land and labour costs, conversion plant sizes, processing technologies and government policies in different regions, results in ethanol production costs and prices that vary considerably by region.

In Brazil and the US, there are a number of large-scale ethanol conversion plants that are considered to be "state-of-the-art". In Europe, there are far fewer plants, and most of them are relatively small and have not been optimised with respect to crops and other inputs. As a result, the typical cost for ethanol produced in Europe is significantly higher than in the US. Brazil is the lowest-cost producer, thanks to lower input costs, relatively large and efficient plants and the inherent advantages of using sugar cane as feedstock.

The IEA Implementing Agreement on Bioenergy recently analysed typical costs for recent large ethanol conversion plants (constructed in the late 1990s) in the US (IEA, 2000a). The average production cost for such plants is estimated to be $0.29 per litre, or $0.43 per gasoline-equivalent litre (Table 4.1). In comparison, the typical refinery "gate price" for gasoline is between $0.18 and $0.25 per litre, depending on world oil prices and other factors.

The largest ethanol cost component is the plant feedstock, although about half of this cost is offset by selling co-products such as "distillers dried grains soluble" (an animal feed). Operating costs represent about one-third of total

Table 4.1

Estimated Corn-to-Ethanol Costs in the US for Recent Large Plants

	Cost per litre
Net feedstock cost	$0.13
Feedstock cost	$0.23
Co-product credit	($0.11)
Operating cost	$0.11
Labour/administration/maintenance	$0.05
Chemical cost	$0.03
Energy cost	$0.04
Capital recovery (per litre)	$0.05
Total production cost (gate price)	$0.29
Cost per gasoline-equivalent litre	$0.43

Source: IEA (2000a). Units are in year 2000 US dollars per litre.

cost per litre, of which the energy needed to run the conversion facility is an important (and in some cases quite variable) component. Capital cost recovery represents about one-sixth of total cost per litre.

An analysis by Whims (2002) shows that plant size has a major effect on cost. Whims estimates that for both dry-mill and wet-mill operations in the US, about a tripling of plant size (from 55 to 150 million litres per year for dry-mill plants and 110 to 375 million litres per year for wet-mill plants) results in a reduction in capital costs of about 40% per unit of capacity, saving about $0.03 per litre. This tripling of plant size can also reduce operating costs by 15% to 20%, saving another $0.02 to $0.03 per litre. Thus, a large plant with production costs of $0.29 per litre may be saving $0.05 to $0.06 per litre over a smaller plant.

The size distribution of ethanol conversion plants in the US as of 1999 is shown in Figure 4.1. The distribution is fairly even, although in recent years most new plants have a capacity greater than 100 million litres per year, contributing to a general decline in average ethanol production cost. According to the Renewable Fuels Association, ethanol plants under construction during 2002 in the US (nearly two billion litres annual capacity) have an average annual production capacity of about 150 million litres (RFA, 1999).

Figure 4.1

US Ethanol Production Plants by Plant Size, as of 1999

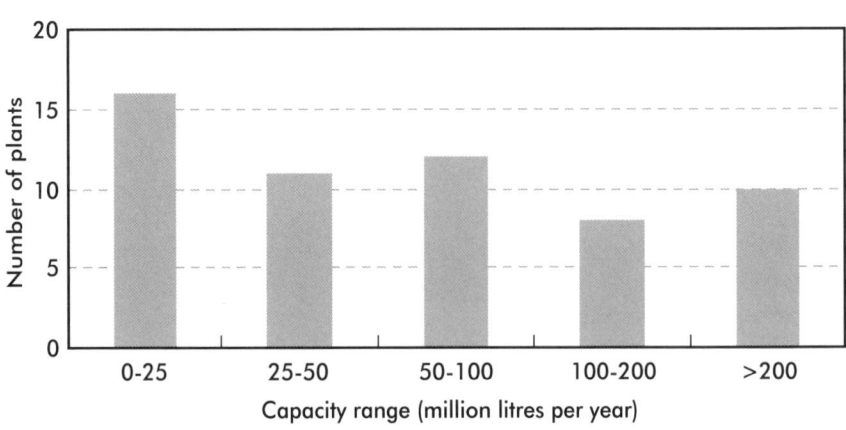

Source: RFA (1999).

Fuel ethanol "gate" prices in the US actually have not declined on average in nominal dollars over the past ten years (Figure 4.2), despite the increasing size of plants and lower costs. The slight downward trend in ethanol prices in the late 1990s was reversed in 2000, but prices have declined again since a peak in 2001. The price of ethanol has averaged about $0.30 per litre over the past twelve years.

Price fluctuation is standard in most commodity markets and in part reflects changes in supply and demand. One important reason for the fluctuation in US ethanol prices is the volatility in feedstock crop prices. Corn prices have varied substantially over the past 20 years, and routinely vary by 50% over any five-year period (Figure 4.2). The price spike for corn in 1995-96 is followed the next year by a price spike in ethanol, although the price spike in ethanol in 2000-2001 actually preceded the smaller corn price spike, reflecting that other market factors can also play an important role in determining prices. In fact, oil prices rose substantially in 2000, which may have helped trigger an increase in ethanol prices.

Probably due to the small number of fuel ethanol plants in Europe and other IEA regions outside North America, few cost studies are available. Table 4.2 provides available estimates for plants in Europe, taken from studies for the IEA's Bioenergy Implementing Agreement and the EU. Costs in Europe are

Figure 4.2

Average US Ethanol and Corn Prices, 1990-2002
(current, unadjusted US dollars)

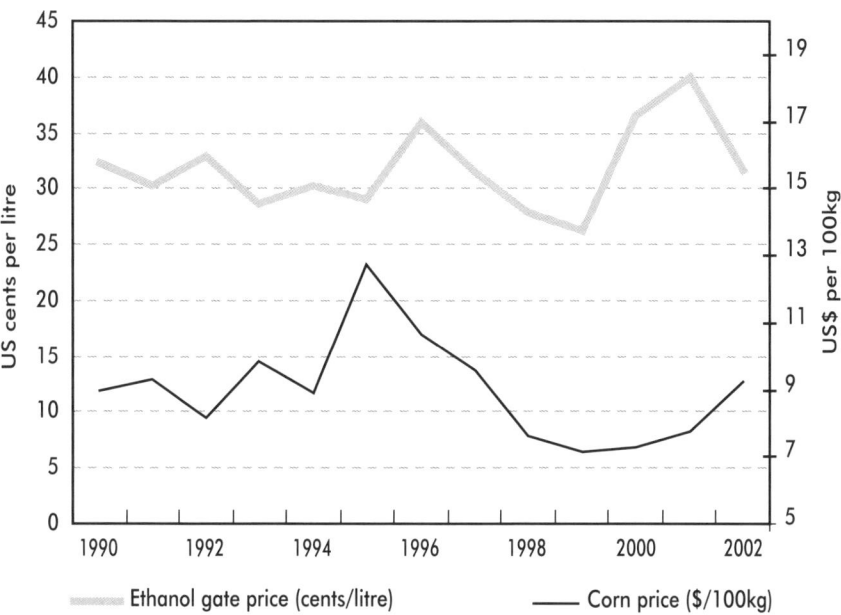

Source: Data for ethanol through 1999 from IEA (2000a); data for 2000-2002 from Whims (2002). Prices are wholesale (plant gate), taxes and subsidies excluded. Data for corn from USDA (2003).

Table 4.2

Ethanol Cost Estimates for Europe
(2000 US dollars per litre)

	Sugar beet	Wheat
Capital costs		$0.08 - $0.13
Feedstock costs		
Raw material	$0.20 - $0.32	$0.22 - $0.34
Co-product credit	($0.00 - $0.01)	($0.11 - $0.15)
Operating costs		$0.20 - $0.25
Other	$0.22 - $0.28	
Total production cost per litre	**$0.42 - $0.60**	**$0.35 - $0.62**
Total per gasoline-equivalent litre	**$0.63 - $0.90**	**$0.53 - $0.93**

Note: For sugar beet, "other" includes all non-feedstock costs.
Sources: IEA (2000b), IEA (2000c), EC-JRC (2002).

much higher than in the US, owing to smaller and less optimised conversion plants, as well as somewhat higher feedstock prices in Europe.

F.O. Lichts (2003) provides some additional insight as to why European ethanol production is more expensive than production in North America. Lichts has compared, on an engineering (theoretical) basis, the relative cost of producing fuel ethanol in the US and in Germany (Table 4.3). Estimates for Germany are for a medium and large-scale plant, operating on wheat and sugar beet. Estimates for the US are for a similar conversion plant using corn. The cost breakdown for the US plant is similar to the estimates of actual US plants in Table 4.1. The engineering estimates for the German conversion plants show that wheat-to-ethanol production is slightly cheaper than sugar beet-to-ethanol. The estimates also reveal significant cost reductions associated with larger plants, about on par with the scale economies described above for US plants. Higher feedstock and energy costs account for the higher overall costs in Germany.

Table 4.3

Engineering Cost Estimates for Bioethanol Plants in Germany, and Comparison to US (US dollars per litre)

Plant capacity	Germany				US	Cost difference (case a minus case e)
	50 million litres		200 million litres		53 million litres	
Raw material	wheat	beet	wheat	beet	corn	
	a	b	c	d	e	
Feedstock cost	$0.28	$0.35	$0.28	$0.35	$0.21	$0.07
Co-product credit	-$0.07	-$0.07	-$0.07	-$0.07	-$0.07	$0.00
Net feedstock cost	$0.21	$0.28	$0.21	$0.28	$0.14	$0.07
Labour cost	$0.04	$0.04	$0.01	$0.01	$0.03	$0.01
Other operating and energy costs	$0.20	$0.18	$0.20	$0.17	$0.11	$0.09
Capital cost recovery (net investment cost)	$0.10	$0.10	$0.06	$0.06	$0.04	$0.06
Total	$0.55	$0.59	$0.48	$0.52	$0.32	$0.23
Total per gasoline-equivalent litre	$0.81	$0.88	$0.71	$0.77	$0.48	$0.34

Source: F.O. Lichts (2003).

Average crop prices in the EU for soft wheat, maize (corn) and sugar beet are shown in Figure 4.3. There are many caveats to a comparison of US crop prices with European prices[1], but nonetheless the comparison can provide a broad indication of the relative feedstock input costs in the two different regions. The average EU price of corn has declined over the past decade, but was still much higher than the US corn price in 2001.

Figure 4.3

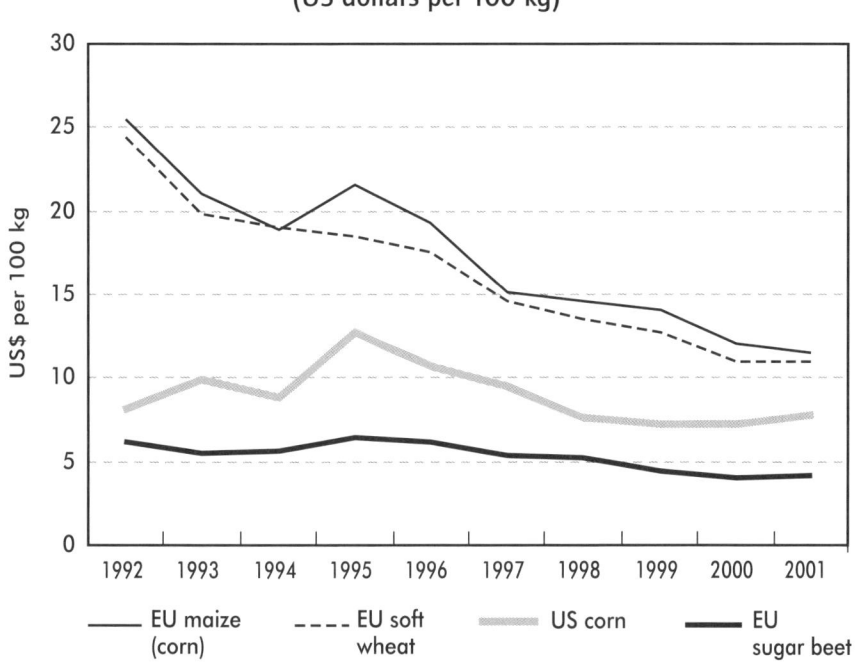

US and EU Average Crop Prices, 1992-2001
(US dollars per 100 kg)

Note: EU data converted to US dollars using nominal yearly average exchange rates.
Sources: USDA / Eurostat.

One important cost factor that is not considered in most of the studies done to date is the presence of subsidies in some of the factors of production, particularly agriculture. Novem/Ecofys (2003) reviews several studies of agricultural subsidies in the EU, and estimates that crop prices might be substantially below their cost of production, and therefore below their "true"

---

1. *Prices per unit crop weight do not fully reflect the costs of using these crops as feedstocks for producing biofuels (e.g. energy per unit weight varies; the conversion efficiency possible with different crops varies; pre-processing costs may vary, crops used at individual plants may vary in terms of quality and price, etc.). All prices are nominal; the EU crop data have been converted to US dollars using nominal exchange rates.*

market price (if there were no agricultural subsidies). The study estimates that the actual cost to produce a litre of ethanol in the EU is about 18 cents higher than reflected in cost estimates that accept crop market prices rather than crop production costs as an input. No similar estimate has been found for the US or other IEA region, but agricultural subsidies of various forms exist in most IEA countries. Subsidies may also exist in the form of tax breaks and incentives to construct conversion plants and other capital equipment. More research is needed to better "weed out" all taxes and subsidies in the development of biofuels cost estimates (the same can be said for most other products, including oil products). The section later in the chapter summarising biofuels costs reflects the possibility of higher real production costs, hidden by agricultural subsidies, in the range of cost estimates provided.

Cost of Ethanol from Sugar Cane in Brazil

Ethanol from sugar cane, produced mainly in developing countries with warm climates, is generally much cheaper to produce than ethanol from grain or sugar beet in IEA countries. For this reason, in countries like Brazil and India, where sugar cane is produced in substantial volumes, sugar cane-based ethanol is becoming an increasingly cost-effective alternative to petroleum fuels. In recent years in Brazil, the retail price (excluding taxes) of hydrous ethanol (used in dedicated ethanol vehicles) has dropped below the price of gasoline on a volumetric basis. In some months in 2002 and 2003, it was even cheaper on an energy basis. Anhydrous ethanol, blendable with gasoline, is still somewhat more expensive. Prices in India have declined and are approaching the price of gasoline. Ethanol prices reached a low in 2002, in part due to over-capacity, and they have risen somewhat in 2003 (IBS, 2003). No other country produces enough cane-derived fuel ethanol to have, as yet, established a clear cost or price regime.

These developments suggest that cane ethanol prices, like prices for many commodities, differ substantially from production costs, and are affected by supply and demand for both ethanol and sugar. For example, government-mandated blending programmes in both Brazil (at 20%-25% per litre of gasoline) and India (5%) have at times driven up prices when ethanol producers had difficulty meeting the demand that these rules required. Flexible-fuel vehicles in Brazil (that can run on any combination of E20/25-blended gasoline and pure (hydrous) ethanol have begun to be sold in Brazil during 2003[2]. As the number of vehicles that can switch, on a daily basis, between E20/25-blended gasoline and pure ethanol increases, gasoline and ethanol prices may become more closely linked.

The only available detailed cost breakdown for ethanol production is from 1990 (Table 4.4). Even then, estimated production costs were lower than much more recent cost estimates for grain and sugar beet ethanol in the US or the EU. Moreover, ethanol production costs in Brazil have declined since 1990.

Table 4.4

Ethanol Production Costs in Brazil, circa 1990

	Average cost (1990 US$ per litre)
Operating costs	$0.167
Labour	$0.006
Maintenance	$0.004
Chemicals	$0.002
Energy	$0.002
Other	$0.004
Interest payments on working capital	$0.022
Feedstock (cane)	$0.127
Fixed costs	$0.062
Capital at 12% depreciation rate	$0.051
Other	$0.011
Total	$0.23
Total per gasoline-equivalent litre	$0.34

Source: C&T Brazil (2002).

Ethanol production costs in Brazil in 1990 were far less than current production costs in the US and the EU[3]. Over the past decade, there have been substantial efficiency improvements in cane production and ethanol conversion processes and more widespread adoption of electricity co-generation using excess cane (bagasse). Both trends have lowered production costs further. As a result, government-set prices paid to producers dropped by 15% between 1990 and 1995, though market prices continue to fluctuate

2. *These vehicles are capable of running on a mix of the types of fuel already available at the pump – E20/25-blended ethanol and pure hydrous ethanol (95% ethanol, 5% water). They use specially designed engines and control systems that can manage such mixtures.*

3. *The figures in Table 4.4 are in 1990 dollars, so they would be considerably higher in 2002 dollars. On the other hand, the Brazilian currency (the real) has dropped in value by over 50% during this period, offsetting much of the inflation over the period.*

dramatically from year to year. Recent production cost estimates for hydrous ethanol are as low as R$ 0.45 per litre, equivalent to US$ 0.15 (at the prevailing exchange rate in January 2004), or $0.23 per gasoline-equivalent litre (USDA, 2003b). Costs for anhydrous ethanol (for blending with gasoline) are several cents per litre higher. When expressed in US dollars, cost estimates are also subject to the considerable fluctuations in exchange rates.

The steady reduction in the cost of producing ethanol in Brazil over time has been masked to some degree by the tumultuous history of the Proalcool programme. Initially, under the programme, the government subsidised ethanol production by paying producers the difference between their production cost and the price they received from distributors (pegged to 25% below the price of gasoline) (Laydner, 2003). When gasoline prices collapsed in the mid-1980s, the subsidy became a heavy burden on the government's budget. In the late 1980s, world sugar prices were high and distilleries drastically cut back on ethanol production in favour of increasing sugar production. This eventually led to a collapse of the entire ethanol programme. The programme was restructured in the mid-1990s, along with removing the oil sector monopoly enjoyed by the national oil company, Petrobras (which had also served as an agent for buying and selling ethanol in some areas). Gasoline and ethanol prices were liberalised "at the pump" in 1996, but ethanol production levels and distillery gate prices were still regulated. A transition to full liberalisation of alcohol prices took place between 1996 and 2000.

As a result of liberalisation, ethanol prices are now driven by market forces and can fluctuate dramatically, influenced in part by world sugar demand and in part by each producer's decision as to how much ethanol to produce. Ethanol prices fell in the late 1990s but recovered by 1999. As shown in Figure 4.4 (showing ethanol distillery "gate" prices compared to refinery gate prices for gasoline), ethanol prices fell again in 2002, recovered in early 2003, and have since fallen to all-time lows[4]. While the government no longer directly subsidises ethanol production, certain tax incentives still exist, including lower taxes on alcohol fuel than on gasoline, lower taxes on the purchase of dedicated ethanol vehicles, and financial incentives to distilleries to encourage them to hold larger alcohol inventories. Including these incentives, the retail price of ethanol as of late 2003 was only about two-thirds that of gasoline on an energy-equivalent

---

4. *Prices in early 2004 (not shown in the figure) were as low as US$ 0.10 (US$ 0.15 per gasoline-equivalent litre), due to a glut of ethanol on the market.*

Figure 4.4

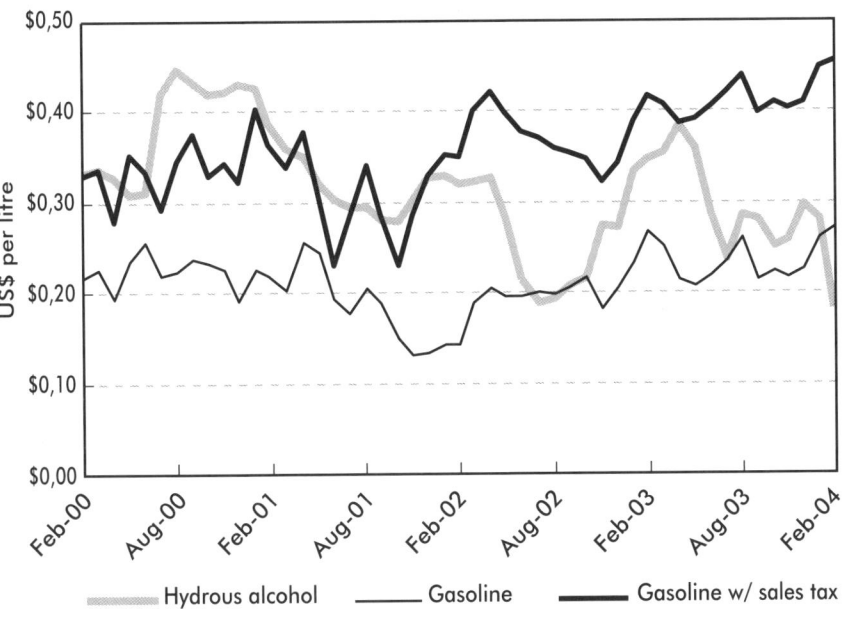

Prices for Ethanol and Gasoline in Brazil, 2000-2003
(US dollars per gasoline-equivalent litre)

Source: Laydner (2003).

basis. Even without the sales-tax advantage (also shown in Figure 4.4), ethanol is now close to competitive with gasoline on a price-per-unit-energy basis at oil prices above $25/barrel (Laydner, 2003).

Ethanol from Cellulosic Feedstock

Ethanol derived from cellulosic feedstock using enzymatic hydrolysis (described in Chapter 2) requires much greater processing than from starch or sugar-based feedstock, but feedstock costs for grasses and trees are generally lower than for grain and sugar crops. If targeted reductions in conversion costs can be achieved, the total cost of producing cellulosic ethanol in OECD countries could fall below that of grain ethanol.

Table 4.5 shows estimates of capital and production costs of cellulosic ethanol from poplar trees from a 2001 assessment of biofuels in the US and Canada. The study was undertaken for the IEA Implementing Agreement on Bioenergy (IEA, 2000a) and the estimates are based on a comprehensive assessment

Table 4.5

Cellulosic Ethanol Plant Cost Estimates
(US dollars per litre except where indicated)

	Near-term base case	Near-term "best Industry" case	Post-2010
Plant capital recovery cost	$0.177	$0.139	$0.073
Raw material processing capacity (tonnes per day)	2 000	2 000	2 000
Ethanol yield (litres per tonne)	283	316	466
Ethanol production (million litres per year)	198	221	326
Total capital cost (million US$)	$234	$205	$159
Operating cost	$0.182	$0.152	$0.112
Feedstock cost	$0.097	$0.087	$0.059
Co-product credit	($0.019)	$0.029	$0.0
Chemicals	$0.049	$0.049	$0.028
Labour	$0.013	$0.011	$0.008
Maintenance	$0.024	$0.019	$0.010
Insurance & taxes	$0.018	$0.015	$0.007
Total cost per litre	$0.36	$0.29	$0.19
Total cost per gasoline-equivalent litre	$0.53	$0.43	$0.27

Source: NREL estimates as quoted in IEA (2000a).

undertaken by the US National Renewable Energy Laboratory in 1999. They are engineering estimates for a large-scale plant with the best available technology, using assumptions regarding technology improvements and cost reductions over the next decade. There are no large-scale commercial cellulosic ethanol plants currently in operation, so it is uncertain whether and when these estimated costs can be achieved in practice. The first large-scale plant is planned for 2006 (EESI, 2003), but it is unlikely that this first plant will achieve these cost-reduction targets.

In Table 4.5, the "near-term base case" cost of $0.36 per litre (about $1.36 per US gallon), as an engineering estimate for a large-scale facility, is only about 20% higher than current production costs for grain ethanol in the United States (Table 4.1). The higher costs for cellulosic ethanol production are mainly due to higher capital recovery costs (conversion plant cost) and to relatively higher operating costs. The cost of the cellulosic feedstock is low compared with grain feedstock costs for conventional ethanol[5].

5. At a feedstock cost of $0.10 per litre of ethanol produced, a fairly large amount of cellulosic feedstock could be economic, at least in the US (see Chapter 6 for a discussion of cellulosic feedstock production potential at various feedstock prices).

A second set of estimates for 2002, NREL's "best industry case", reflects improvements that might occur if several large plants were built and optimised ("n^{th} plant"). In this case, overall production cost is $0.29 per litre, about 20% lower than the $0.36 estimate for the near-term base case, and about equal to the current industry cost for producing ethanol from corn. Table 4.5 also provides estimates of future costs based on potential technical advances. NREL estimates that costs could drop to as low as $0.19 per litre in the post-2010 time frame, due to lower plant construction costs and improved conversion efficiency. If so, cellulosic ethanol would probably become cheaper than grain ethanol.

Can such cost reductions be achieved over the next 10-15 years? The US National Research Council, in a recent report reviewing the US biofuels research programme (NRC, 1999), expressed concern that the optimistic cost estimates made over the past decade have not yet been realised. The report was somewhat sceptical about whether estimates such as those in Table 4.5 can be achieved in this time frame. However, the US Department of Energy has recently refocused its cellulosic ethanol research programme on new areas of potential cost reduction. As mentioned in Chapter 2, several new types of test facilities will be constructed within the next several years.

Assuming that over the next decade a number of large-scale commercial plants are built, and cellulosic ethanol production costs experience the hoped-for decline, Table 4.6 compares possible cellulosic ethanol costs with projected costs for ethanol from corn and with gasoline prices in the US. The prices in Table 4.6 exclude existing fuel taxes and subsidies. Costs for corn and cellulosic ethanol are in terms of gasoline-equivalent litres (and thus are about

Table 4.6

Gasoline and Ethanol: Comparison of Current and Potential Production Costs in North America (US dollars per gasoline-equivalent litre)

	2002	2010	Post-2010
Gasoline	$0.21	$0.23	$0.25
Ethanol from corn	$0.43	$0.40	$0.37
Ethanol from cellulose (poplar)	$0.53	$0.43	$0.27

Notes and sources: Gasoline gate cost based on $24/barrel oil in 2002, $30/barrel in 2020; corn ethanol from IEA (2000a), with about 1% per year cost reduction in future; cellulosic costs from IEA (2000a) based on NREL estimates.

50% higher than for an actual litre of ethanol). Production costs for corn ethanol are assumed to decline slowly, and feedstock prices are assumed to remain roughly constant in real terms.

Based on the energy-equivalent cost and price projections in Table 4.6, ethanol produced in OECD countries will not likely compete with gasoline before 2010. After 2010, however, cellulosic ethanol could compete, if targets are met. Of course, all types of ethanol have a better chance of becoming competitive, and sooner if oil prices are above the assumed $24/barrel. Research on biofuels is driven in part by increasing concern about the mid to long-term decline of petroleum production and the upward pressure this trend could eventually have on gasoline and diesel prices. Under this scenario, marginal production cost decreases for ethanol will be greatly assisted by pump price increases in conventional fuels.

At least one study has estimated the cost of producing cellulosic ethanol in Europe (Novem/ADL, 1999). This study's cost estimate for cellulosic ethanol appears to be quite similar to the estimate from NREL. This study is reviewed in the section on advanced processes, below.

Biodiesel Production Costs in the United States and the European Union

Biodiesel production costs are even more dependent on feedstock prices than are ethanol costs. Recent work undertaken for the IEA Bioenergy Implementing Agreement reviewed production cost at six European biodiesel facilities, and provides a range of cost estimates (Table 4.7). As with ethanol,

Table 4.7

Biodiesel Cost Estimates for Europe
(US dollars per diesel-equivalent litre)

Scenario	Rapeseed oil price	Conversion costs	Final cost
Small scale, high raw material price	0.60	0.20	0.80
Small scale, low raw material price	0.30	0.20	0.50
Large scale, high raw material price	0.60	0.05	0.65
Large scale, low raw material price	0.30	0.05	0.35

Source: IEA (2000d), with conversion to diesel-equivalent litres.

production scale has a significant impact on cost, but since this is a smaller share of overall cost, it is less significant than for ethanol. In general, costs for production via "continuous process" are lower than for "batch" processes. The range of cost estimates shown in Table 4.7 is for production from rapeseed oil in the EU, where most of the world's biodiesel is produced.

It is important to note that these cost estimates include, as a credit, the value of co-product sales (as is also true for the ethanol estimates above). However, glycerine is a key co-product for biodiesel, and glycerine markets are limited. Under a scenario of large-scale production of biodiesel, the excess supply of glycerine (or "glycerol") could cause its price to fall to near zero. Glycerine prices in Europe currently range from $500 to $1 000 per tonne, depending on quality; this figure varies substantially depending on supply availability (USDA, 2003). Since glycerine is produced at a ratio of 1:10 with methyl ester, the co-product credit of glycerine is on the order of $0.05-$0.10 per litre of biodiesel produced. This improves the economics of biodiesel production significantly, and the costs in Table 4.7 would increase by up to $0.10 if glycerine prices collapsed.

In the US, fewer large-scale production facilities exist, and costs appear to be slightly higher. US biodiesel production relies mainly on soy oil, which is generally more available, and lower-priced in the US, than rapeseed oil. Coltrain (2002) estimates US biodiesel production costs ranging from about $0.48 to $0.73 per diesel-equivalent litre. This range is based on soy oil costs of $0.38 to $0.55 per litre of biodiesel produced, production costs in the range of $0.20 to $0.28 per litre, and a glycerol credit of about $0.10 per litre. These estimates are consistent with other recent studies (*e.g.* ODE, 2003).

Thus, the current cost of producing biodiesel from rapeseed ranges from $0.35 to $0.65 for large-scale facilities in the EU and perhaps $0.10 more at the smaller-scale plants in the US, per conventional diesel-equivalent litre of biodiesel (taking into account that biodiesel has about 10% less energy per litre than petroleum diesel fuel). This figure could rise by an additional $0.10 under large-scale production, if it caused the price of glycerine to fall. Gate prices for petroleum diesel typically range between $0.17 and $0.23 per litre, depending on world oil price.

Costs are lower for biodiesel produced from waste greases and oils, since the feedstock price is lower. But quantities of biodiesel from these waste sources are generally limited – although organised collection practices could significantly increase their availability. Costs for biodiesel from waste greases and oils can be as low as $0.25 per litre, in cases where the feedstock is free (or even, in a few cases, where the feedstock has a negative price – where companies are willing to pay to have it removed from their sites). However, such low-cost greases and oils (such as "trap grease") typically also are impure and need additional processing before conversion into methyl ester (Wiltsee, 1998). Further, production with waste feedstocks often occurs at a very small scale, which can increase capital and operating costs. Thus, the amount of biodiesel that could be produced at very low cost may be quite small, relative to diesel fuel use in most IEA countries. Costs on the order of $0.30 to $0.40 per litre may be more typical (ODE, 2003).

Although the cost of biodiesel production can be expected to decline somewhat as larger-scale plants are built with further design optimisations, there appear to be few opportunities for technical breakthroughs that would lead to substantial cost reductions in the future. The cost of the feedstock is the dominant factor. Biodiesel could be cheaper to produce in countries with lower cost oil-seed crop prices, typical around the developing world.

Production Costs for Advanced Biofuels Production Technologies

As discussed in Chapter 2, there are a variety of other methods under investigation for producing liquid and gaseous fuels from biomass feedstock. Available estimates of production costs for most processes, including gasification, Fischer-Tropsch synthesis, "biocrude" liquefaction and other approaches, are based mainly on engineering studies and associated estimates of potential cost reductions. Most advanced processes appear expensive, and the potential for future cost reductions is uncertain. This section focuses on a review of one recent, comprehensive study that provides cost estimates for a wide range of processes (Novem/ADL, 1999). The costs in Table 4.8 correspond to the processes described in Chapter 2 and to the oil/greenhouse gas impacts discussed in Chapter 3. They are based on the same study and assumptions as in Table 3.6.

Table 4.8 shows the results of Novem/ADL's cost assessment for a variety of biomass gasification and other biomass-to-liquids (BTL) processes. The estimates are based on the assumption that costs are reduced through scale and technology improvements. For example, ethanol from biomass is shown at

Table 4.8

Estimates of Production Cost for Advanced Processes

Fuel	Feedstock / location	Process	$/litre gasoline-equivalent
Diesel	petroleum	refining	$0.22
Biodiesel	rapeseed	oil to FAME (transesterification)	$0.80
Diesel	biomass - eucalyptus (Baltic)	HTU	$0.56
Diesel	biomass - eucalyptus (Baltic)	gasification / F-T	$0.68
Diesel	biomass - eucalyptus (Baltic)	pyrolysis	$1.36
DME	biomass - eucalyptus (Baltic)	gasification / DME conversion	$0.47
Gasoline	petroleum	refining	$0.22
Ethanol	biomass - poplar (Baltic)	enzymatic hydrolysis (CBP)	$0.27
Ethanol	biomass - poplar (Brazil)	enzymatic hydrolysis (CBP)	$0.27
Gasoline	biomass - eucalyptus (Baltic)	gasification / F-T	$0.76
Hydrogen	biomass - eucalyptus (Baltic)	gasification	$4.91
CNG	biomass - eucalyptus (Netherlands)	gasification	$0.46

Note: Average gate prices for gasoline and diesel in 1999 in the Netherlands are also shown.
Source: Novem/ADL (1999).

$0.27 per gasoline-equivalent litre, similar to the cost estimated by NREL for production in North America in the post-2010 time frame (Table 4.5). Most of the estimates are quite high, ranging from $0.47 per litre for dimethyl ether (DME) to nearly $5.00 per litre for hydrogen from biomass gasification. These do include all production and distribution costs to make these fuels available at retail stations. A substantial portion of hydrogen costs relate to developing a fuel distribution and refuelling infrastructure.

Some of the processes are close to competing with conventional biofuels, *e.g.* biomass-to-diesel fuel using hydrothermal upgrading (HTU – the "biocrude" process), and can nearly compete with biodiesel from crop oils in some regions. Research on these advanced processes is ongoing and technical breakthroughs beyond those considered in the Novem/ADL study could well occur. If costs can be reduced to acceptable levels, they could become very attractive options for future transport fuels, given their high conversion efficiencies and very low well-to-wheels greenhouse gas emissions.

Summary of Biofuels Production Costs

As the preceding sections have shown, biofuels production costs can vary widely by feedstock, conversion process, scale of production and region. Though only a few specific studies have been presented, the "point estimates" that have been shown should not be misinterpreted as indicating low variability. Based on these estimates and a more general review of the literature, the following Figures 4.5 and 4.6 provide IEA's best estimates regarding near-term and long-term ranges of biofuels' production costs. The rather wide ranges provided also reflect the possibility that some hidden costs, such as agricultural subsidies, are not fully reflected in the reviewed studies[6].

On an energy basis, ethanol is currently more expensive to produce than gasoline in all regions considered (Figure 4.5). Only ethanol produced in Brazil comes close to competing with gasoline. Ethanol produced from corn in the US is considerably more expensive than from sugar cane in Brazil, and ethanol from grain and sugar beet in Europe is even more expensive. These differences reflect many factors, such as scale, process efficiency, feedstock costs, capital and labour costs, co-product accounting, and the nature of the estimates (for example, all available estimates for cellulosic ethanol are engineering-based, rather than from actual experience).

Considerable cost reductions are possible over the coming decade and beyond, as shown in the "post-2010" section of Figure 4.5. If costs reach the low end of the ranges depicted here, ethanol in all regions will be more cost-competitive than it is currently. In particular, the cost of both ethanol from sugar cane in Brazil (and probably in many other developing countries) and cellulosic ethanol in all regions of the world have the potential to reach parity or near-parity with the cost of gasoline, with oil prices between $25 and $35 per barrel. Gasification of biomass, with Fischer-Tropsch synthesis to produce synthetic gasoline, is not expected to be competitive in the next 10-15 years, unless unanticipated breakthroughs occur.

Figure 4.6 shows cost ranges for diesel fuel and different biodiesel replacement options. Biodiesel from rapeseed in the EU appears to be somewhat more competitive with diesel than ethanol is with gasoline; however, in the US biodiesel is generally farther from competitive prices than

---

6. *Following Novem/Ecofys (2003), the high end of the cost range for each fuel/feedstock combination has been extended somewhat to reflect possible price impacts from excluding agricultural subsidies.*

Figure 4.5

Cost Ranges for Current and Future Ethanol Production (US dollars per gasoline-equivalent litre)

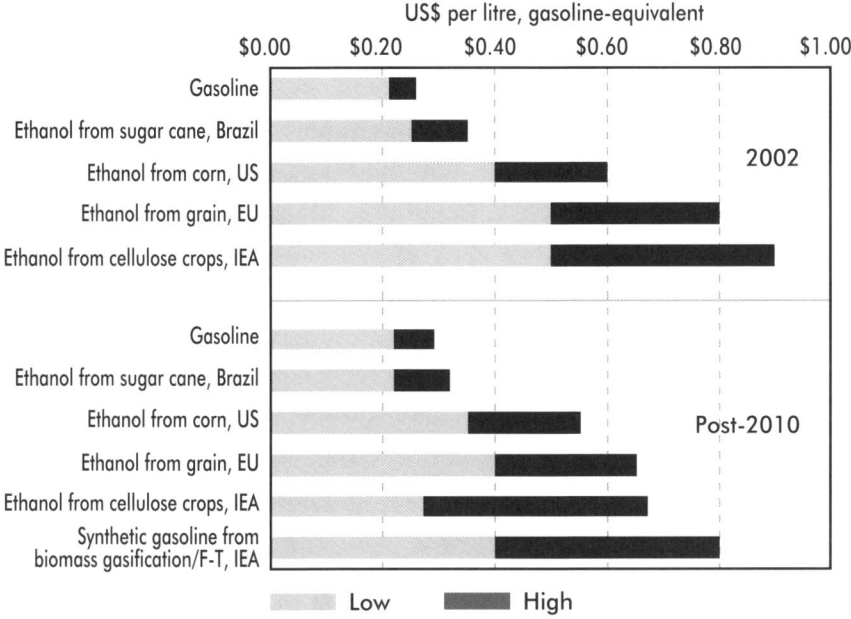

Source: IEA analysis.

is ethanol. In both regions, biodiesel production costs show a wide range for many of the same reasons mentioned above for ethanol. The value of biodiesel co-products helps to bring the net production cost under $0.50 per litre in the US, and below $0.40 in the EU, at the lower end of the cost spectrum. For biodiesel from waste feedstocks (like yellow grease), costs can nearly compete with diesel, though only for certain situations and at relatively small volumes.

In the longer term, biodiesel costs may, on average, not change significantly from their current levels. For biodiesel from FAME, any cost reductions from scale and improved technology could easily be offset by higher crop prices and/or a decline in the value of co-products like glycerine. However, new types of biodiesel, such as hydrothermal upgrading (HTU) and biomass gasification followed by Fischer-Tropsch conversion to synthetic diesel, could compete with other forms of biodiesel. But none of these types of biodiesel is expected to match the cost of conventional diesel fuel, at least if they are produced in IEA

Figure 4.6

Cost Ranges for Current and Future Biodiesel Production
(US dollars per litre, petroleum-diesel-fuel-equivalent)

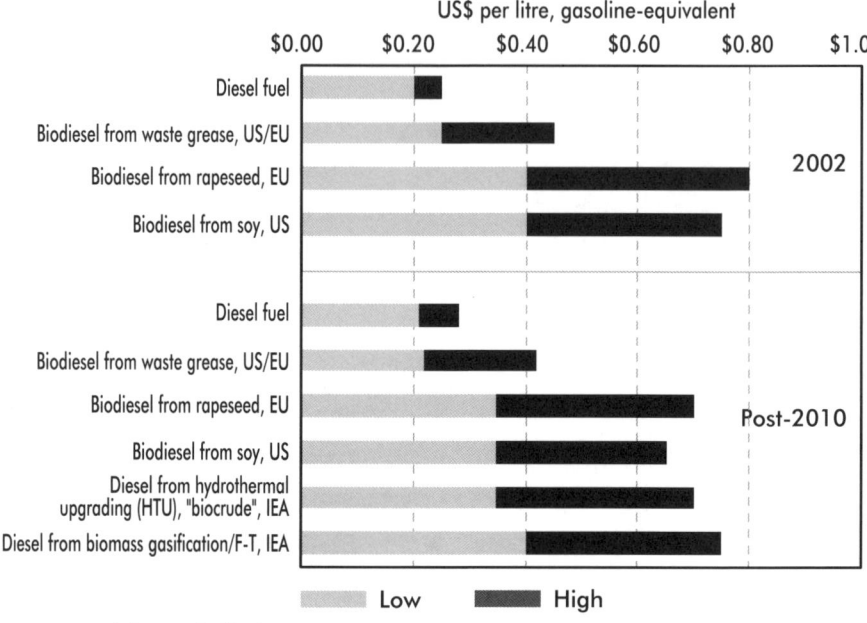

Source: IEA analysis. Note: "F-T" is Fischer-Tropsch type process.

countries. This study has not looked closely at the possibility of biodiesel fuel produced in developing countries, as few data are available. But it is likely that any of the new types of biodiesel could be made more cheaply in developing countries than in IEA countries.

Biofuels Distribution and Retailing Costs

Although biofuels production costs are by far the biggest component of retail prices (apart from possibly large taxes/subsidies in some countries), distribution and retailing costs can be significant when biofuels must be transported over land great distances to reach markets. Biodiesel fuel is much easier to transport than ethanol, because it can use the same transport and storage infrastructure as conventional diesel. Similarly, equipment that is used to store, transport and deliver diesel can also be used for biodiesel with no modification, whereas minor modifications and some degree of fuel separation is needed for ethanol. Since biodiesel from FAME is non-toxic, it does not require additional safety measures for storage and handling or special training of full service attendants. The main costs of transporting biodiesel are the cost of shipping it from the production facility to the storage terminal and the cost of storing it before blending with petroleum diesel (EC, 1998). Low production volumes can increase per unit costs considerably. Since ethanol is an alcohol, it creates some compatibility problems with the existing infrastructure, which are not applicable to biodiesel. As such, the following discussion focuses on these issues and the additional costs associated with using ethanol.

Ethanol Transportation Costs

The ethanol distribution chain begins at the production facility, where 100% ethanol is denatured, for example by blending it in 5% gasoline. The blend is typically shipped from the plant to a bulk storage terminal for redistribution by tanker truck, rail car or river barge. The product is stored at the terminal until sufficient supplies have been collected for distribution on to fuelling stations. Final blending with gasoline typically occurs at the terminal, when the gasoline delivery truck is loaded in preparation for delivery to fuelling stations (known as "splash" blending). The cost estimates here are based largely on the US, where most data are available. Most of the cost estimates are based on shipping ethanol from the Midwest to California, as the volume along this route is expected to increase dramatically (see Chapter 7).

Tanker trucks normally deliver ethanol to markets located close to ethanol production plants. This can cost as little as a few US cents per gallon or less than one cent per litre (DA, 2000). Since the cost of producing and transporting ethanol is the primary limitation to widespread use, the largest ethanol fuel markets in the US and Europe have emerged close to feedstock areas and production facilities.

An important consideration in estimating shipping costs is scale. Rail, pipeline and shipping are only viable options on a fairly large scale, with bulk movement of liquid fuel. Rail transport is cost-effective for shipping ethanol more than several hundred kilometres. In the absence of pipelines, rail will probably be the preferred transport mode to move ethanol from the Midwest in the US to the large markets on the east and west coasts. Production plants with less than 250 million litres per year capacity might require a bulk storage terminal. The terminal could store output from several plants, thus boosting volume and making rail transport more economic. Rail transit from Midwest ethanol plants to California takes from two to three weeks, and costs $0.03 to $0.05 per litre, depending on the plant of origin and the market destination (DiPardo, 2002). With a large enough market, dedicated 100-car, or "unit", trains could cut costs considerably.

If available, the cheapest mode of transportation to many markets is via pipeline. Ethanol and ethanol-gasoline blends are not currently shipped by pipeline owing to a number of technical and operational difficulties. Primary among these is ethanol's sensitivity to water. Many countries (including the US) use a "wet" pipeline system, which means that unless moisture is removed, ethanol could absorb water and arrive at its destination off specification. Most pipeline segments would also need to undergo some type of preparatory cleaning to remove built up lacquers and other deposits to prevent contamination of the ethanol and trailing products in the system. Additionally, in the US most pipelines originate in the Gulf Coast and run north, northeast and northwest. Since most ethanol feedstock and production plants are located in the Midwest, it would be necessary to barge the product south to access many pipeline markets. Construction of dedicated pipelines for transportation of ethanol or gasoline-ethanol blends is not currently viewed as feasible with the current low shipment volumes.

Table 4.9

Ethanol Transportation Cost Estimates for the US

Mode/distance	Price range per litre
Water (including ocean and river barge)[a]	$0.01 to $0.03
Short trucking (less than 300 km)[b]	$0.01 to $0.02
Long-distance trucking (more than 300 km)	$0.02 to $0.10
Rail (more than 500 km)[c]	$0.02 to $0.05

Note: The distribution cost estimates are based on data from the Midwest US.
Sources: [a] Forum (2000), [b] DA (2002), [c] DiPardo (2002).

Storage / Distribution Terminal Costs

Ethanol is normally stored, and finally blended with gasoline, at product terminals. In order for an existing terminal to initiate an ethanol blending programme it must have a tank of sufficient size to meet projected ethanol demand. The tank or tanks must also be large enough to receive the minimum shipment size while still maintaining adequate working inventory. Blending systems must be installed (or existing blending systems modified) to accommodate gasoline-ethanol blending. "Splash blending" is sometimes used, where ethanol is mixed with gasoline as the tanker truck is being filled; however, this may result in incomplete blending and higher volatility of the product.

The estimated cost of installing a 25 thousand barrel tank is about US$ 500 000, while costs for blending systems and modifications to receive ethanol at the terminal could push costs up to $1 million. However, with reasonably well utilised equipment (for example 24 tank refills per year), the costs would only be $0.002 per litre of ethanol stored (DA, 2000).

Refuelling Stations

Gasoline refuelling pumps can easily be adjusted to accommodate ethanol, either as a blend with gasoline or as pure ethanol (EU-DGRD, 2001). Low percentage ethanol blends, such as E10, are currently dispensed in service stations in many countries with few reported problems. Ethanol blends with a higher alcohol concentration, such as E85, however, have a tendency to degrade some materials, and they require minor modifications or replacement of soft metals such as zinc, brass, lead (Pb) and aluminium. Terne (lead-tin alloy)-plated steel, which is commonly used for gasoline fuel tanks, and lead-based solder are also incompatible with E85. Non-metallic materials like natural rubber, polyurethane, cork gasket material, leather, polyvinyl chloride (PVC), polyamides, methyl-methacrylate plastics, and certain thermo and thermoset plastics also degrade when in contact with fuel ethanol over time. If these materials are present, refuelling station storage and dispensing equipment may need to be upgraded or replaced with ethanol-compatible materials such as unplated steel, stainless steel, black iron and bronze. Non-metallic materials that have been successfully used for transferring and storing ethanol include non-metallic thermoset reinforced fibreglass, thermo plastic piping, neoprene rubber, polypropylene, nitrile, Vitorn and Teflon materials. The best choice for underground piping is non-metallic corrosion-free pipe (NREL, 2002).

There are other essential steps for refuelling station conversion. The tank (or liner) material must be compatible with gasoline-ethanol blends and any water-encroachment problems must be eliminated. Materials and components in submersible pumps must be compatible with gasoline-ethanol blends. Tanks, especially older tanks, should be cleaned.

Independent retailers estimate the typical cost for converting one retail unit with three underground storage tanks in the US to be under $1 000 (DA, 2000). For complete replacement of tanks or pumps, or for a new installation, costs are much greater. The cost of adding a 3 000 gallon (11 400 litre) E85 tank and accessories in Kentucky was about $22 000 (NREL, 2002) as shown in Table 4.10. The costs of retail conversion for E10-compatibility are small, typically less than $0.002 per litre on a per unit basis. New E85 retail station infrastructure is more expensive, possibly exceeding 2 cents per litre of ethanol.

Overall, the total cost of transporting, storing and dispensing ethanol ranges from about $0.01 to $0.07 per litre (Table 4.11). These cost estimates are based on the foregoing discussion, which was based mainly on US data. However, only some of these costs can be considered as incremental to the cost of gasoline – which also must be moved, stored and dispensed. Thus in the long run, if ethanol capacity expansion occurs *instead* of gasoline capacity expansion, then the incremental cost of moving, storing and retailing ethanol might be fairly small, probably in the lower half of the range indicated in Table 4.11.

Table 4.10

Cost of Installing Ethanol Refuelling Equipment at a US Station

	Cost
Cost of 3 000 gallon storage tank and accessories	$16 007
Dispensing equipment (all alcohol-compatible), including: • Single hose pump • 1 micron fuel filter • Alcohol whip hose • 8 feet of pump hose • Breakaway valves • Swivel hose • Fuel nozzle • Anti-siphon valve	$3 400
Cost to offload tank	$440
Tank connections and internal plumbing	$454
Wire system and programme to existing fleet management system operated by Mammoth Cave National Park, Kentucky	$1 915
Total cost of project	**$22 216**

Source: NREL (2002).

Table 4.11

Total Transport, Storing, and Dispensing Costs for Ethanol (US dollars per litre)

	Cost range per litre
Shipping cost	$0.010 to $0.050
Storage/blending cost	$0.000 to $0.002
Dispensing cost	$0.002 to $0.020
Total cost	$0.012 to $0.072

Source: Data presented in this chapter.

Biofuels Cost per Tonne of Greenhouse Gas Reduction

In simplest terms, the cost of using biofuels to reduce greenhouse gas emissions, by substituting for oil use in vehicles, depends on just two factors: the net (generally well-to-wheels) greenhouse gas reduction per litre or per kilometre, and the incremental cost of the fuel used (per litre or per kilometre)[7]. Figure 4.7 shows the potential range of cost per tonne of greenhouse gas reduction from biofuels. Their incremental cost, per energy-equivalent litre, over gasoline is plotted on the vertical axis, and the percentage well-to-wheels GHG emissions reduction is on the horizontal axis. The lines in the figure represent "isocosts", *i.e.* along each line, the cost per tonne is constant.

As discussed in the preceding sections of this chapter, the incremental cost per litre of biofuel (in gasoline-equivalent litres) ranges from 0 to about $0.50, and estimates of the percentage reduction in well-to-wheels greenhouse gases per litre of biofuels range from 0 to about 100%. Therefore, the cost per tonne can range from zero up to $500 per tonne or more (though $500 is the highest value line shown in the figure)[8]. The greenhouse gas control strategies currently being considered by IEA countries are generally less than $50 per tonne of CO_2, so for biofuels to be an attractive option, their incremental cost must be fairly low and their greenhouse gas percentage reduction fairly high. Thus, the

---

7. As discussed in Chapter 8, many non-cost or difficult-to-quantify aspects make it much more difficult to estimate a full social cost of biofuels.

8. At $0.50 incremental cost and 10% GHG reduction, the cost per tonne is $1 750.

Figure 4.7

Cost per Tonne of CO$_2$ Reduction from Biofuels in Varying Situations

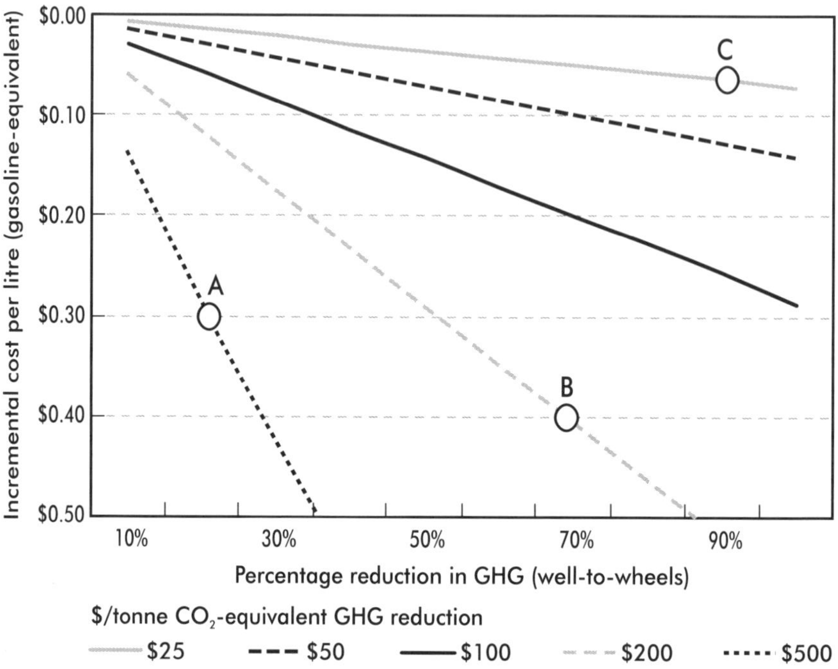

$/tonne CO$_2$-equivalent GHG reduction

——— $25   – – – $50   ——— $100   – – – $200   ••••• $500

top two lines in the figure ($25 and $50 per tonne isocost lines) represent the most likely cases where biofuels would be attractive to policy-makers.

Three points have been added to the figure to illustrate the resulting cost-per-tonne estimates for different combinations of biofuel cost and greenhouse gas reduction, in a hypothetical case. At point A, with an incremental cost of $0.30 per litre and GHG reductions of 20%, the cost per tonne is about $500. However, if the use of biofuels could cut GHG emissions by much more, say 70%, then its cost per tonne could be cut significantly, even if the per-litre cost of the fuel were higher. This case is shown at point B. Finally, if biofuels with very high GHG reduction potential, say 90%, could be produced at a modest incremental cost per litre, e.g. $0.15, then (as shown at point C) the cost-per-tonne reduction would drop to about $50, competitive with many other policy options.

Figure 4.8 provides a range of cost-per-tonne estimates for various biofuels, currently and post-2010, based on the estimates of CO$_2$ reduction in Chapter 3 and the cost estimates developed in this chapter. As shown, cost-per-tonne estimates vary considerably, with ethanol from grain crops in the US and EU providing among the highest-cost greenhouse gas reductions, at least given current costs and reduction characteristics. This is due to the fact that, at the high end, ethanol may not provide substantial GHG reductions, and its incremental cost may be quite high – similar to point A in the previous figure. However, over time, cost reductions and ongoing improvements in emissions characteristics should eventually bring the cost per tonne from grain ethanol to within the $200-$400 range. In contrast, cellulosic ethanol could already provide GHG reductions at a cost per tonne of $300 or less, if large-scale plants were constructed, and this cost could come down to under $100 after 2010 (and probably after several large-scale plants have been built). This difference in cost per tonne suggests that supporting the use of a relatively expensive fuel that provides large emissions reductions, like cellulosic ethanol, may be worthwhile.

Figure 4.8

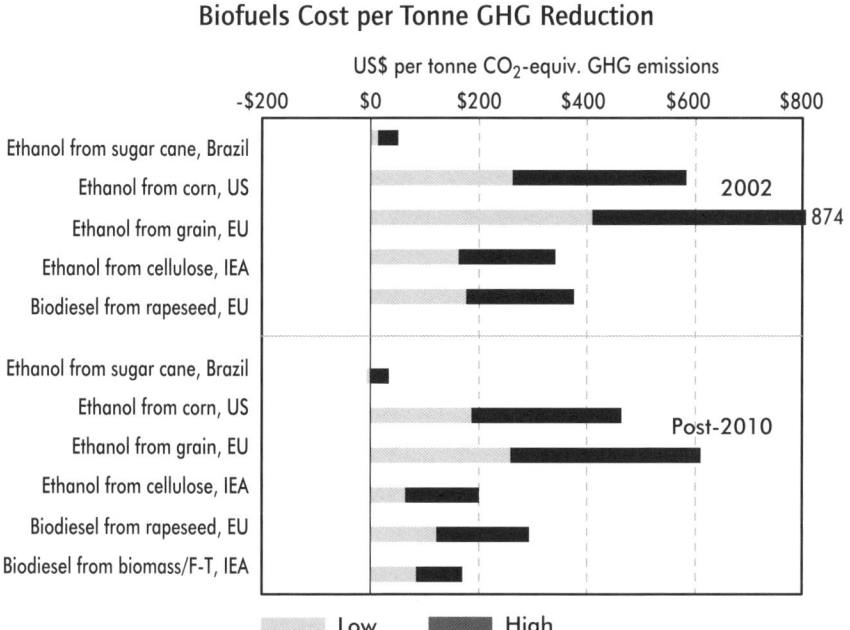

Biofuels Cost per Tonne GHG Reduction

Note: Ranges were developed using highest cost/lowest GHG reduction estimate, and lowest cost/highest GHG reduction estimate for each option, then taking the 25% and 75% percentile of this range to represent the low and high estimates in this figure. In some cases, ranges were developed around point estimates to reflect uncertainty. Source: Cost data are from tables in this chapter. GHG reduction data are from Chapter 3.

In Brazil, ethanol production already appears to be cost-competitive with other GHG reduction options, with an estimated range of about $20-$60 per tonne. Brazil is operating in the vicinity of point C (or better) in Figure 4.7. Eventually, Brazilian costs will likely drop to $0 per tonne, or even become negative, if ethanol becomes cheaper to produce than gasoline on an energy-equivalent basis.

Though biodiesel from oil-seed crops (FAME) are quite expensive to produce, they can outperform grain ethanol in terms of GHG reduction, and thus have a significantly lower cost per tonne, especially at the high end of the range. However, over the longer term, biodiesel has less potential for cost reduction than ethanol, particularly cellulosic ethanol.

Finally, new technologies for producing biofuels, such as biodiesel from hydrothermal upgrading (HTU) or from biomass gasification with Fischer-Tropsch (F-T) synthesis, could provide GHG reductions at under $100 per tonne within the next 10-15 years. This reflects still quite high costs of production with these processes, but their very high emissions reductions yield a relatively low cost per tonne. Since these relatively new processes are still under development, their potential for long-term cost reduction, even beyond that shown in the figure, appears good.

The cost-per-tonne estimates in Figure 4.7 only take into account the direct production and distribution costs, not the various other costs and benefits of biofuels, such as energy security, pollutant emissions reductions and octane enhancement. If these additional attributes were taken into account, the net cost per tonne of GHG reduction could be much lower. These attributes are discussed further in Chapter 8, though given the difficulty in quantifying them, no attempt is made to develop "social cost" estimates of biofuels. It is an important area for future research.

Crop Market Impacts of Biofuels Production

Total biofuel costs should also include a component representing the impact of biofuels production on related markets, such as food. As crops (or cropland) are drawn away from other uses, prices can rise. In Brazil, a large switch away from producing sugar to producing ethanol could affect the price of sugar around the world. Estimating these "market equilibrium" impacts is not simple.

The relationship between markets and products is complex and is related to even greater complexities in the macroeconomy as well as to specific policies that affect the relevant markets. Therefore, the relationship tends to be different in different countries and over time.

Despite these difficulties, several recent studies have attempted to quantify the broader market impacts of producing biofuels in various markets and regions. A study by Raneses *et al.* (1999) examined the potential impacts of increased biodiesel use in the US. The authors looked at potential demand for biodiesel fuel in three markets: federal fleets, mining and marine/estuary travel. These markets combined are estimated to have the potential to consume nearly 400 million litres per year. This is less than 1% of current US motor diesel fuel use, but more than current biodiesel consumption. The authors calibrated an agricultural model (Food and Agricultural Policy Simulator, FAPSIM) and tested the impact of changes in biodiesel demand, through its soy oil requirements, on the market price of soy oil and other agricultural commodities (as soybeans compete for land and displace other crops). Biodiesel production can also affect the production of soybean meal, a co-product of biodiesel production and a livestock feed. As production of biodiesel increases, so does production of this co-product, causing its price to decline.

The results of the simulations for low, medium and high biodiesel demand are summarised in Table 4.12. As soybean oil demand rises, its price and the price of other crops rise, while the price of soybean meal falls. In the high scenario (with greater biodiesel demand from the three target markets), soybean oil production rises by 1.6%, driving up soybean oil price by 14.1% (an elasticity of over 8), reflecting relatively inelastic supply. The soybean meal price drops by 3.3%, reflecting excess supply relative to demand. Soybean crop prices rise by 2%, slightly more than unitary elasticity with respect to the production of soybean oil (about one litre of biodiesel can be made from one litre of soybean oil). This is a fairly substantial change in the price of soybeans, again reflecting somewhat inelastic supply. Livestock prices drop by 1.4% reflecting lower feed costs, while US farm income increases by 0.3%.

A similar type of simulation modelling effort was undertaken by Walsh *et al.* (2002). This study evaluated the potential market impacts of growing switchgrass, poplar and willow for the production of cellulosic ethanol. Their analysis shows that not only can increased demand for certain crops lead to

Table 4.12

**Estimated Impacts from Increased Use of Biodiesel
(Soy Methyl Ester) in the US**

	Market scenario (percentage change from baseline)		
	Low	**Medium**	**High**
Soybean oil production	0.3	0.8	1.6
Soybean oil price	2.8	7.2	14.1
Soybean meal price	-0.7	-1.7	-3.3
Soybean price	0.4	1.0	2.0
Livestock price ("broilers")	-0.3	-0.7	-1.4
US net farm income	0.1	0.2	0.3

Source: Raneses *et al.* (1999).

an increase in the price of those crops, but it can also increase the price of other crops competing for the same agricultural land.

Taking into account potential lands for production of the dedicated energy crops, some of which are also croplands used to produce food and fibre crops, the authors calibrated a crop model to investigate the following:

■ The price for cellulosic crops necessary to bring them into production on different types of land.

■ The likely regional distribution of energy crops brought into production.

■ The potential impacts on traditional crop production and prices.

■ The impact on US farm income.

■ The economic potential for using a modified US Conservation Reserve Program to serve as a source of bioenergy crops.

The scenarios were summarised by two "bounding cases". The principal results from these two cases are shown in Table 4.13.

Clearly, the study indicates that production of bioenergy crops would compete with traditional cropland in the US and could lead to higher crop prices. In the two scenarios shown, crop prices rise anywhere from 4% to 14%, depending on the scenario and crop. This would essentially lead to a transfer from consumers to farmers and from urban areas to rural areas. Given the objective

Table 4.13

Estimated Impacts from Increased Production of Switchgrass for Cellulosic Ethanol on Various Crop Prices

	Case I	Case II	Notes
Farm gate price of switchgrass (per dry tonne)	$33	$44	Higher switchgrass price draws more into production
Total land brought into production of energy crops (million hectares)	7.1	17	Primarily from switchgrass
Reduction in land allocated to current crops (million hectares)	4.2	9.5	Switched to bioenergy crops (remainder of bioenergy crops grown on marginal and set-aside lands)
Change in price of selected traditional crops			From reduction in production of these crops
Corn	4%	9%	
Sorghum	5%	14%	
Wheat	4%	12%	
Soybeans	5%	10%	
Cotton	9%	13%	
Rice	8%	10%	

Source: Walsh *et al.* (2002).

of many countries to maintain and improve farm incomes and rural communities, these may be desirable impacts. If not, some of these impacts presumably could be avoided if bioenergy crops were restricted to being grown on non-crop land (such as marginal and conservation reserve lands). In these scenarios, this would cut the amount of bioenergy crops available by nearly two-thirds.

A third study, by Koizumi (2003), looked at the Brazilian ethanol programme and its impacts on world ethanol and sugar markets. This is an important case study due to the size and potential interactions of the markets for these two products. Brazilian sugar producers have a major impact on world sugar prices through their decisions on how much sugar to produce. These decisions, in turn, are related to their weighing the relative profitability of producing sugar or producing alcohol with the cane resources they process. Most cane processors in Brazil have considerable flexibility in producing different combinations of refined sugar and ethanol. They make decisions based on domestic and international prices and in turn can have a great impact on these prices.

Using the ethanol/sugar market model, the author built several scenarios through 2010 testing the impact of different levels of ethanol demand on ethanol production, sugar production and international sugar prices. Under a scenario with strong ethanol production, where the "allocation ratio" for sugar production decreases from 48% to 45% (a 3% reduction in share, but a 6% reduction in sugar production), the domestic sugar price would increase by up to 28%. World sugar prices are estimated to increase by up to 4%. Thus, the model suggests that there would be a significant wealth transfer from sugar-importing countries to Brazil and other sugar-exporting countries.

Government policies can have a significant and unforeseen impact on the way in which markets react to changes in production levels. Ugarte and Walsh (2002) assessed the potential impact of a US policy to encourage production of switchgrass beginning in 1996, the year of a major change in agricultural support policies. In that year, the US replaced farm supports based on suppressing crop production with direct payments that were unrelated to production levels, though with safeguards if prices fell below certain levels. Over the following four years, overproduction led to a decline in prices below the specified levels which triggered substantial payments, including some emergency interventions to help prevent large-scale bankruptcies in the farming sector.

The authors, using the POLYSIS agricultural model, estimate that if, instead of the policy adopted, the US had encouraged production of switchgrass as a new energy crop that did not compete directly with food crops, the prices of food crops would not have dropped as much as they did. The results suggest that the government could have saved up to $2 billion in net agricultural subsidies while total farm incomes would have been higher (with revenues from selling switchgrass). Thus, the net effect of subsidising switchgrass production would have been to more than offset other subsidies, while increasing farm incomes. Many countries have extensive systems to support farmers. Ugarte and Walsh's analysis indicates that the introduction of dedicated energy crops could reduce other existing subsidy costs – such as in situations where farmers receive support if their incomes fall below certain levels. Thus, although growing bioenergy crops might in some cases compete for land with other crops and increase crops prices, on the other hand it may help to redirect existing farming subsidies to more productive purposes.

These studies show that the market impacts of changes in the production of particular fuels and particular crops can be complex and far-reaching, and agricultural policies may have unexpected impacts. Much more analysis is clearly needed in this area, and greater efforts should be made by policy-makers to account for full economic equilibrium effects of new biofuels-related policies.

Macroeconomic Impacts of Biofuels Production

While increased crop demand may trigger an increase in crop prices, as well as in other related markets, there are also important potential "macro" benefits from increasing the domestic production of biofuels. Sims (2003) points out that the benefits of oil displacement include the positive contribution to a country's balance of trade and domestic economic activity. Brazil reduced its oil import bill by an estimated $33 billion between 1976 and 1996 through the development of its ethanol industry. The full benefits are difficult to measure, requiring general equilibrium modelling and assumptions regarding the costs and risks of oil import dependence, such as the risk of supply disruption or sudden spikes in prices.

Biofuels production in developing countries can also have a positive impact on agricultural labour employment and rural development, particularly when conversion facilities are smaller-scale and are located near crop sources in rural districts. In Brazil for example, it is estimated that 700 000 jobs have been created in rural areas to support the additional sugar cane and bioethanol industry. The development of multi-product "biorefineries" could further spur the development of related secondary industries.

In addition to employment benefits, domestic biofuels production enhances the security of national energy supply and improves the balance of trade, since many countries spend large percentages of their foreign currency reserves on oil imports.

The potential economic benefits from developing biofuels must be weighed against the costs of producing the biofuels, and the negative economic impact these higher costs have on government budgets and economic growth. Such effects must be carefully assessed before the broader macroeconomic benefits are used as justification for biofuels production.

---

5 VEHICLE PERFORMANCE, POLLUTANT EMISSIONS AND OTHER ENVIRONMENTAL EFFECTS

Biofuel costs, and the impact of their use on oil demand and greenhouse gas emissions are important components in the overall assessment of biofuels for transport. But there are other factors, such as the impact of biofuels on the vehicles that use them and the pollutant emissions from these vehicles, which are also relevant to this analysis. This chapter explores some of these additional impacts.

Vehicle-Fuel Compatibility

Biofuels have the potential to leapfrog a number of traditional barriers to entry faced by other alternative fuels because they are liquid fuels, largely compatible with current vehicles and blendable with current fuels. Moreover, they can share the long-established gasoline and diesel motor fuel distribution infrastructure, in many cases with little required modification to equipment. Low-percentage ethanol blends, such as E5 and E10, are already dispensed in many service stations worldwide, with almost no reported incompatibility with materials and equipment. Biodiesel from fatty acid methyl esters (FAME) is generally accepted to be fully blendable with conventional diesel, except for certain considerations when using high-percentage biodiesel blends or neat (pure) biodiesel. Another type of biodiesel, synthetic diesel fuel produced from biomass gasification and Fischer-Tropsch synthesis, is even closer in composition to conventional diesel fuel and blendability is a non-issue.

Ethanol Blending in Gasoline Vehicles

Efforts to introduce ethanol into the transport fuel market has, in most countries, focused on low-percentage blends, such as ethanol E10, a 10% ethanol to 90% gasoline volumetric blend (sometimes known as "gasohol"). More recently, research and road tests have examined higher-percentage ethanol blends and pure (neat) alcohol fuels, and have focused on the modifications that must be made to conventional gasoline vehicles in order to use these blends. The United States has been particularly active in its research and testing of these blends (Halvorsen, 1998).

Ethanol and Materials Compatibility

What are the potential problems with operating conventional gasoline vehicles with an alcohol-gasoline blend? Alcohols tend to degrade some types of plastic, rubber and other elastomer components, and, since alcohol is more conductive than gasoline, it accelerates corrosion of certain metals such as aluminium, brass, zinc and lead (Pb). The resulting degradation can damage ignition and fuel system components like fuel injectors and fuel pressure regulators (Otte et al., 2000).

As the ethanol concentration of a fuel increases, so does its corrosive effect. When a vehicle is operated on higher concentrations of ethanol, materials that would not normally be affected by gasoline or E10 may degrade in the presence of the more concentrated alcohol. In particular, the swelling and embrittlement of rubber fuel lines and o-rings can, over time, lead to component failure.

These problems can be eliminated by using compatible materials, such as Teflon or highly fluorinated elastomers (Vitorns) (EU-DGRD, 2001). Corrosion can be avoided by using some stainless steel components, such as fuel filters. The cost of making vehicles fully compatible with E10 is estimated to be a few dollars per vehicle. To produce vehicles capable of running on E85 may cost a few hundred dollars per vehicle. In the US, however, several car models capable of operating on fuel from 0% to 85% alcohol are sold as standard equipment, with no price premium over comparable models.

It is widely accepted in the literature, as well as by the fuels and car manufacturer communities, that nearly all recent-model conventional gasoline vehicles produced for international sale are fully compatible with 10% ethanol blends (E10). These vehicles require no modifications or engine adjustments to run on E10, and operating on it will not violate most manufacturers' warranties (EU-DGRD, 2001; Novem/Ecofys, 2003[1]). However, many vehicle operators may not be aware of this high degree of compatibility and concerns about using this fuel blend are still common. One legitimate source of concern is with older models – many manufacturers have increased the ethanol compatibility of their vehicles in recent years (*e.g.* during the 1990s) and in some countries a higher share of older models still on the road may not be fully compatible with ethanol blends like E10.

Low-level ethanol blends (E5 and especially E10) are widely used in the US, Canada, Australia and in many European countries, where they have delivered over a trillion kilometres of driving without demonstrating any significant differences in operability or reliability (AAA, 2002; Forum, 2000). E10 typically has a slightly higher octane than standard gasoline and burns more slowly and at a cooler temperature. It also has higher oxygen content and burns more completely, which results in reduced emissions of some pollutants, as discussed below.

In blend levels above E10, some engine modifications may be necessary, though the exact level at which modifications are needed varies with local conditions such as climate, altitude and driver performance criteria (EAIP, 2001). In Brazil, cars with electronic fuel injection, including imported cars built for the Brazilian market with minor modifications (such as tuning and the use of ethanol-resistant elastomers), have operated satisfactorily on a 20% to 25% ethanol blend since 1994. There have been few reported complaints about drivability or corrosion (Moreira, 2003).

In the US, limited research has shown that conventional, unmodified gasoline vehicles also appear capable of operating on ethanol blends that are higher than 10%. In a study on the effects of ethanol blending in cold climates, the Minnesota Center for Automotive Research (MnCAR) examined vehicle operations on ethanol blends up to 30% by volume (MSU, 1999). The project tested fifteen standard, unmodified light-duty vehicles, fuelled with E10 and E30 and operated under normal driving conditions, over a period of one year. MnCAR examined fuel economy, emissions, drivability and component compatibility characteristics. The study revealed no drivability or material compatibility problems with any of the fifteen vehicles tested (though long-

---

1. *The Novem/Ecofys study lists ethanol blend-level warrantees for a sample of recent models from major manufacturers. These range from E5 for Volkswagen models to E15 for most Renault models. Most models are warranted at least to E10.*

term effects were not tested for). However, other studies have indicated that ethanol blends have relatively poor hot-fuel handling performance, due to high vapour pressure. Fuel formulation to control vapour pressure is necessary to ensure smooth running in warm climates and higher altitudes (Beard, 2001).

Fuel economy can be re-optimised for higher ethanol blends through minor vehicle modifications. Given ethanol's very high octane, vehicles expected to run on ethanol blend levels of over 10% (such as in Brazil) can be re-optimised by adjusting engine timing and increasing compression ratio, which allows them to run more efficiently on the higher blend levels, and saves fuel. On an energy basis, a 20% blend of ethanol could use several percentage points less fuel with a re-optimised engine. Some newer vehicles automatically detect the higher octane provided by higher ethanol blends, and adjust timing automatically. This could result in immediate fuel economy improvements on ethanol blends (taking into account ethanol's lower energy content), but it is not clear just how much fuel economy impact this could have. A number of studies have tested (or reviewed tests of) the fuel economy impacts of low-level ethanol blends (*e.g.* Ragazzi and Nelson, 1999; EPA, 2003; Novem/Ecofys, 2003). These have found a fairly wide range of impacts, from slightly worse to substantially better energy efficiency than the same vehicles on straight gasoline. Tests have typically been conducted on just a few vehicles and under laboratory conditions rather than as actual in-use performance. More research is needed on this very important question.

High-level Ethanol Blends

Following on the successful applications of E10 in several countries and E22-26 in Brazil, considerable interest surrounds the use of much higher-level blends, particularly E85 (85% ethanol, 15% gasoline). In this light, US demand for E85 grew from about 500 000 litres in 1992 to eight million litres in 1998, spurred in part by government requirements on certain vehicle fleets and the availability of credits that can be earned under US fuel economy requirements (DiPardo, 2002). The latter incentive has helped spur several car manufacturers to initiate large-scale production of E85 vehicles, and there are now over two million of these in operation in the US – though few currently run on E85 fuel. In Sweden, a strong push is under way to introduce E85 fuel, with about 40 stations in place as of 2002. However, far fewer E85 vehicles have been sold there, perhaps just a few thousand as of 2002 (NEVC, 2002).

As with low blend E10, E85 vehicles can overcome two of the important barriers faced by most alternative fuel vehicles: incremental vehicle cost and refuelling. The cost of mass-producing flexible-fuel vehicles (FFV) is believed to be some $100 to $200 per vehicle[2] – much lower than the several thousand dollars of incremental cost to produce vehicles running on compressed natural gas, LPG or electricity. And high-level ethanol blends can be distributed through existing refuelling infrastructure with relatively minor changes.

One area that needs addressing in building an FFV is that E85 has a lower vapour pressure than gasoline at low temperatures, which makes it more difficult to create an ignitable vapour in the combustion chamber. E85 must be heated either as it enters or once it is in the fuel rails to improve cold starting and to reduce cold-start emissions (Otte *et al.*, 2000). The ethanol can be heated by an intake manifold heater or an inlet air heater. Most effective cold-starting ethanol vehicles use integrated air heaters and thermal storage systems (Halvorsen, 1998). In addition to heating the fuel droplets, an increase in fuel pressure is needed.

The E85 vehicles sold in large numbers in the US typically have an engine control computer and sensor system that automatically recognises what combination of fuel is being used. The computer also controls the fuel and ignition systems to allow for real-time calibration[3]. Some components in the fuel system, like the fuel tank, filter, pump and injectors, are sized differently and made of material which is compatible with the higher concentration of alcohol and resists corrosion, such as a stainless steel fuel tank and Teflon-lined fuel hoses (ICGA, 2003).

FFVs look and drive like "gasoline only" vehicles, and many car owners may be unaware of their vehicle's ability to operate with E85 fuel. In the US, FFVs can be purchased or leased from new automobile dealerships. Flex-fuel capability is standard equipment on several models and are covered under the same warranty, service and maintenance conditions as their gasoline-powered counterparts (Ford, 2003).

Ethanol-gasoline blends above 85% can pose problems for gasoline engines, but pure or "hydrous" ethanol (actually a mixture of 96% ethanol and 4% water by volume) can be used in specially designed engines. This type of

---

2. *Though recent, objective estimates of incremental production costs of FFVs could not be obtained.*
3. *This involves an increased flow rate of fuel through the injectors and a change in spark plug timing.*

engine has been in use in Brazil for many years. Engines need to be protected against corrosion to be compatible with this fuel, but they do not require the system for identifying and adjusting for different fuel mixtures that FFVs need. They are thus not appreciably more expensive to produce than FFVs. Hydrous ethanol is cheaper to produce than anhydrous ethanol (used for blending with gasoline), since the water that occurs naturally during the production of alcohol must be removed before blending.

Dedicated ethanol vehicles can be re-designed to take full advantage of ethanol's very high octane number. In Brazil, some engine manufacturers have increased vehicle compression to 12:1, compared to the typical 9:1 ratios of conventional gasoline vehicles.

Ethanol Blending with Diesel Fuel

Although ethanol is generally thought of as a blend only for spark-ignition (gasoline) vehicles, it is also possible to blend it into diesel fuel (actually an emulsion, since ethanol is not naturally miscible with diesel fuel). Since ethanol can generally be produced more cheaply than biodiesel, and with greater output per unit land devoted to growing feedstock (as discussed in Chapter 6), its use as a diesel blend could be very interesting in some countries for expanding biofuels use into the diesel fuel "pool". However, until recently its use in diesel has been limited because its low cetane number makes it very difficult to burn by compression ignition. As such, the main research in diesel-ethanol technology has been to find ways to force ethanol to ignite by compression in the diesel engine (Murthy, 2001). Recent developments in the areas of new additives to improve ethanol solubility in diesel, and improve ignition properties, have made ethanol blending into diesel an interesting and viable option.

A number of approaches have been developed to improve ethanol-diesel solubility. One method is to essentially give carburettor benefits to a diesel engine, where the diesel is injected in the normal way, and a carburettor is added to atomise the ethanol into the engine's air stream. Under this "dual fuel" operation, diesel and ethanol are introduced into the cylinder separately (SAE, 2001). A number of comprehensive trials have been carried out in northern Europe to assess the use of "E-diesel", a generic name for an ethanol-diesel blended motor fuel comprised of up to 15% ethanol and up to 5% special additive solubilising emulsifiers (MBEP, 2002). An emulsifier is required, even at 5% ethanol, to prevent the ethanol and diesel from separating at very low temperatures or if water contamination occurs by improving the water tolerance of ethanol-diesel blends. In addition to solubilising effects, a number of other benefits are claimed for the emulsifiers, including improved lubricity, detergency and low-temperature properties. Also, because of the low cetane number of ethanol (and therefore poor auto-ignition properties) the additive package must also include a cetane-enhancing additive such as ethylhexylnitrate or ditertbutyl peroxide (McCormick and Parish, 2001).

Many millions of miles of fleet testing using low-level ethanol-diesel blends have been logged in Europe (Sweden, Denmark, Ireland), Brazil, Australia, and the United States (Nevada, Illinois, Nebraska, Texas and New York City). Sweden has tested a variant of E-diesel for many years in urban buses operating in Stockholm, with great success. Using Swedish Mark II diesel fuel, perhaps the cleanest in the world as the base, this 15% ethanol blend with up to 5% solubiliser has shown significantly improved emissions performance and reliable revenue service. Brazil has also pioneered the investigation of E-diesel since the late 1990s, demonstrating that a properly blended and formulated E-diesel can operate quite successfully in a very warm, humid climate. Generally, the results of US E-diesel fleet testing to date have indicated that a fuel with less than 8% ethanol in most applications, particularly in stop-and-go urban operations, has no adverse effect on fuel efficiency when compared to the performance of "typical" low-sulphur diesel (Rae, 2002).

Another quite different approach, experimented with since the early 1990s, has been to modify diesel engines to adjust their fuel auto-ignition characteristics, in order to be able to run on very high ethanol blends, such as 95% ethanol. Like low-level blends, ethanol use in E95 engines requires an "ignition-improving" additive which helps initiate the combustion of the fuel and decreases the ignition delay, though with the engine modifications, such additives were actually easier to develop, and were developed earlier, than for ethanol in low-level blends in conventional diesel engines.

In 1992, Archer Daniels Midland (ADM) put into service the first fleet of ethanol-powered, heavy-duty trucks for evaluation and demonstration in the US (ADM, 1997). Four trucks were equipped with specially modified Detroit Diesel Corporation model 6V-92TA engines and were fuelled with E95, composed of 95% ethanol and 5% gasoline. Substantial engine modifications were necessary, including to the electronic control module and the electronic fuel injectors. Also, because ethanol contains only about 60% of the energy of diesel fuel per unit volume, more ethanol fuel is required to generate the same amount of power in the engine. Therefore, larger ethanol-resistant fuel pumps were used and the diameter of the holes in the injector tip were increased. The bypass air system was modified to achieve the proper ethanol compression ignition temperature. The ethanol engines also incorporated a glow plug system to help start the engine. Another major modification to the ethanol engine was an increased compression ratio from 18:1 for diesel to 23:1 for E95.

Other tests of high-level ethanol blends in modified diesel engines have now been carried out in Minnesota and Sweden. In both cases, the vehicles performed well, although in the Minnesota trial the maintenance and repair costs of the E95 trucks were considerably higher, primarily due to fuel filter and fuel pump issues. From an emissions standpoint, the E95 trucks appear to emit less particulate matter and fewer oxides of nitrogen but more carbon monoxide and hydrocarbons than their conventional diesel-fuelled counterparts (Hennepin, 1998). The Swedish programme is probably the world's largest, and is ongoing. By 1996, there were roughly 280 Volvo and Scania buses in 15 cities running on neat, 95% ethanol, with an additive to improve ignition. Scania has assisted in developing one blending agent, used to create the ethanol formulation "Beraid", that is now undergoing approval in the European Union as a reference fuel for diesel engines that run on ethanol (RESP, 2003a). Another formulation, called "Puranol", has been developed by the Pure Energy Corporation.

Clearly, ethanol in low- and high-level blends for use in diesel engines is a viable option, but one that deserves further research and attention, particularly in countries where a significant substitution of biofuels for petroleum diesel fuel is sought.

Biodiesel Blending with Diesel Fuel

Biodiesel from fatty acid methyl esters (FAME) is very suitable for use in standard compression-ignition (diesel) engines designed to operate on petroleum-based diesel fuel. Unlike ethanol, biodiesel can be easily used in existing diesel engines in its pure form (B100) or in virtually any blend ratio with conventional diesel fuels. Germany, Austria and Sweden have promoted the use of 100% pure biodiesel in trucks with only minor fuel system modifications; in France, biodiesel is often blended at 5% in standard diesel fuel and at 30% in some fleet applications. In Italy, it is commonly blended at 5% in standard diesel fuel (EU, 2001). In the US, the most common use is for truck fleets, and the most common blend is B20.

Lower-level (*e.g.* 20% or less) biodiesel blends can be used as a direct substitute for diesel fuel in virtually all heavy-duty diesel vehicles without any adjustment to the engine or fuel system (EC, 1998 and NREL, 2000). The use of biodiesel in conventional heavy-duty diesel engines does not appear to void the engine warranties of any major engine manufacturer (though warranty restrictions to 5% biodiesel are common for light-duty vehicles). Tractor-maker John Deere recently modified its formal warranty statement to affirm the company's endorsement of the use of low-blend biodiesel fuels in its equipment, and according to industry sources, the use of biodiesel in conventional diesel engines does not void engine warranties of any of the major engine manufacturers (Lockart, 2002).

Since pure FAME biodiesel acts as a mild solvent, B100 is not compatible with certain types of elastomers and natural rubber compounds and can degrade them over time (NREL, 2001). However, with the trend towards lower-sulphur diesel fuel, many vehicle manufacturers have constructed engines with gaskets and seals that are generally biodiesel-resistant. On the other hand, the solvent properties of biodiesel have also been noted to help keep engines clean and well running. In some cases, standard diesel leaves a deposit in the bottom of fuelling lines, tanks and delivery systems over time. Biodiesel can dissolve this sediment, but the deposits may then build up in the fuel filter. Initially, the filters may need to be changed more frequently with biodiesel. But once the system has been cleaned of the deposits left by the standard diesel, the vehicle will run more efficiently (BAA, 2003a). Biodiesel also cleans the fuel system of waxes and gums left behind by previous diesel use, including unblocking injectors.

As FAME biodiesel ages (*e.g.* sits in an idle vehicle for several weeks), it can begin to degrade and form deposits that can damage fuel injection systems (Ullmann and Bosch, 2002). Therefore, depending on the blend level and the typical use patterns of the vehicle, special considerations may be necessary for long-term operation on biodiesel fuel (NREL, 2001; Fergusson, 2001). In general, the higher the blend level, the more potential for degradation. In particular, the use of B100 may require rubber hoses, seals and gaskets to be replaced with more resistant materials, other non-rubber seals or biodiesel-compatible elastomers. The quality of biodiesel has also been found to be an important factor in its effects on vehicle fuel systems, and standardisation of fuel quality requirements is considered an important step.

Biodiesel mixes well with diesel fuel and stays blended even in the presence of water. Biodiesel blends also improve lubricity. Even 1% blends can improve lubricity by up to 30%, thus reducing engine "wear and tear" and enabling engine components to last longer (NREL, 2000). Therefore, although biodiesel contains only about 90% as much energy as diesel fuel, with its higher burning efficiency (due to the higher cetane number) and its better lubricity, it yields an "effective" energy content which is probably just a few percentage points below diesel. In over 15 million miles of field demonstrations in Australia, biodiesel showed similar fuel consumption, horsepower, torque and haulage rates as conventional diesel fuel (BAA, 2003b). Biodiesel also has a fairly high cetane number (much higher than ethanol, for example), which helps ensure smooth diesel engine operation (see box).

Diesel Fuel and the Cetane Number

Ignition quality in diesel fuel is measured by the "cetane number". The cetane number measures how easily ignition occurs and the smoothness of combustion. To a point, a high number indicates good ignition, easy starting, starting at low temperature, low ignition pressures, and smooth operation with lower knocking characteristics. Low-cetane fuel reflects poor ignition qualities causing misfiring, carnish on pistons, engine deposits, rough operation and higher knocking. The cetane number requirement for an engine depends on the engine design, size, operational speed, load condition and atmospheric conditions. In the US, a typical cetane number range for "#2 diesel fuel" is 40-45 while for #1 diesel it is 48-52. In the EU a minimum of 49 cetane is normally required. Biodiesel from vegetable oils can have a cetane number in the range of 46 to 52, and animal fat-based biodiesel cetane numbers range from 56 to 60 (Midwest, 1994). The cetane number can be improved by adding certain chemical compounds, but some of these increase vehicle pollutants (EC, 1998).

In cold weather, the difficulty of starting a cold engine increases as the fuel cetane number decreases. With slightly lower cetane ratings than petroleum diesel fuel, FAME biodiesel and B20 congeal or "gel" sooner in very cold (below freezing) temperatures. The "cold flow" properties of FAME depend on the type of vegetable oil used – for example, rapeseed and soy methyl ester are better than palm oil methyl ester. Precautions beyond those already employed for petroleum diesel are not needed when using B20, but for B100 certain simple preventative measures are recommended, including utilising a block heater to keep the engine warm; utilising a tank heater to keep fuel warm during driving; keeping the vehicle inside; and blending with a diesel-fuel winterising agent (Tickell, 2000). Moreover, additives are now available that improve biodiesel's ability to start up engines in cold weather (ARS, 2003).

Impacts of Biofuels on Vehicle Pollutant Emissions

When used either in their 100% "neat" form or more commonly as blends with conventional petroleum fuels, biofuels can reduce certain vehicle pollutant emissions which exacerbate air quality problems, particularly in urban areas.

Biofuels (ethanol and FAME biodiesel) generally produce lower tailpipe emissions of carbon monoxide (CO), hydrocarbons (HC), sulphur dioxide (SO_2) and particulate matter than gasoline or conventional diesel fuel, and blending biofuels can help lower these emissions. Ethanol-blended gasoline, however, produces higher evaporative HC (or volatile organic compounds, VOCs). Impacts of both ethanol and biodiesel on oxides of nitrogen (NO_x) are generally minor, and can be either an increase or a decrease depending on conditions. They also can have impacts, some positive and some negative, on toxic air emissions. Biofuels are generally less toxic to handle than petroleum fuels and in some cases they have the additional environmental benefit of reducing waste through recycling. Waste oils and grease can be converted to biodiesel, and cellulosic-rich wastes, which currently inundate landfills, can be converted to ethanol.

The principal petroleum-related, mobile-source emissions are particulate matter (PM), volatile organic compounds (VOCs) and nitrogen oxides (NO_x), carbon monoxide (CO) and a variety of unregulated toxic air pollutants. VOCs and NO_x are precursors for tropospheric ozone. Weather and local geographic characteristics are important factors in determining the impact of these air pollutants; for example, ozone formation occurs more easily in hot weather and CO is a bigger problem in cold weather and at high altitudes. Toxic air pollutants are more pronounced in hot weather (Andress, 2002). Air pollutants can be emitted from motor vehicle systems both by the exhaust system and by evaporation from the fuel storage, an important factor since ethanol has high volatility and generally increases evaporative emissions of gaseous hydrocarbons.

In OECD countries, the ongoing implementation of increasingly strict emissions control standards for cars and trucks will tend to mute the air quality impacts of biofuels, since manufacturers are required to build vehicles that meet these standards under a range of conditions. In many non-OECD countries, however, emissions control standards are less strict and biofuels are likely to have a larger impact on emissions. This will change over time, as in many developing countries new vehicles are increasingly being required to meet basic emissions standards. Worldwide, older vehicles with little or poor quality emissions control equipment can certainly benefit from the use of biofuels, particularly in terms of reductions in emissions of PM and sulphur oxides.

Emissions from Ethanol-Gasoline Blends

As shown in Table 5.1, ethanol blends such as E10 typically reduce emissions of a variety of pollutants relative to gasoline, though increase certain others. Using ethanol instead of MTBE as an oxygenate in reformulated gasoline (RFG) produces a similar range of impacts. However, some of these impacts are much larger or more important than others.

Among the biggest impacts from using ethanol are on reducing carbon monoxide emissions. Use of a 10% ethanol blend in gasoline is widely documented to achieve a 25% or greater reduction in carbon monoxide emissions by increasing the oxygen content and promoting a more complete combustion of the fuel (*e.g.* DOE, 1999; CSU, 2001; EPA, 2002a). The use of ethanol as a fuel "oxygenate" in parts of the US and in some other countries is mainly for this purpose.

The net impacts of using ethanol on emissions of volatile organic compounds (VOCs) and oxides of nitrogen (NO_x), which combine in the atmosphere to form ozone, are less clear. When ethanol is added to gasoline, evaporative

Table 5.1

Changes in Emissions when Ethanol is Blended
with Conventional Gasoline and RFG

	Ethanol-blended gasoline vs. conventional gasoline	Ethanol-blended RFG vs. RFG with MTBE
Commonly regulated air pollutants		
CO	–	–
NO_x	+	n.c.
Tailpipe VOC	–	n.c.
Evaporative VOC	+	n.c.
Total VOC	+	n.c.
Particulate matter	–	–
Toxic/other air pollutants		
Acetaldehyde	+	+[b]
Benzene	–	–
1,3 Butadiene	–	–
Formaldehyde	+[a]	–
PAN	+	+[b]
Isobutene	–	–
Toluene	–	–
Xylene	–	–

Notes: Minus (–) used for decrease in emissions, plus (+) used for increase. "n.c.": no change.
[a] Formaldehyde emissions decrease for ethanol blends compared with MTBE blends.
[b] A California study concluded that the ambient air concentrations of acetaldehyde and PAN (peroxyacetyl nitrates) increased only slightly for California RFG3 containing ethanol, despite the fact that the increase in primary acetaldehyde emissions is significant. The study concluded that most of the increase in acetaldehyde and PAN concentrations were due to secondary emissions. No comparable study has been done for federal RFG for areas outside California.
Source: ORNL (2000).

VOCs can increase due to the higher vapour pressure, measured as Reid Vapour Pressure (RVP), of the ethanol mixture. Generally, adding the first few per cent of ethanol triggers the biggest increase in volatility; raising the ethanol concentration further does not lead to significant further increases (and in fact leads to slight decreases), so blends of 2%, 5%, 10% and more have a similar impact. In most IEA countries, VOC emissions and thus RVP must be controlled in order to meet emissions standards. In order to maintain an acceptable RVP, refiners typically lower the RVP of gasoline that is blended with ethanol by reducing lighter fractions and using other additives (ORNL,

2000). In Canada, regulations require that the volatility of ethanol blends must at least match that of standard gasoline (CRFA, 2003).

Use of low-level ethanol blends usually does not markedly change the level of nitrogen oxide emissions relative to standard gasoline. Evidence suggests that NO_x levels from low-level ethanol blends range from a 10% decrease to a 5% increase over emissions from gasoline (EPA, 2002a; Andress, 2002). However, over the "full fuel cycle", which takes into account emissions released during ethanol production, feedstock production and fuel preparation, NO_x emissions can be significantly higher. This is primarily due to NO_x released from the fertiliser used to grow bioenergy crops, and occurs mostly outside urban areas.

Emissions of most toxic air pollutants decrease when ethanol is added to gasoline, primarily due to dilution of gasoline, which emits them. Emissions of acetaldehyde, formaldehyde and peroxyacetyl nitrate (PAN), however, increase when ethanol is added. But toxic emissions of benzene, 1,3-butadiene, toluene and xylene, all of which are considered more dangerous, decrease with the addition of ethanol. Formaldehyde and acetaldehyde, like particulate matter, are not present in fuel but are by-products of incomplete combustion. They are formed through a secondary process when other mobile-source pollutants undergo chemical reactions in the atmosphere. PAN, which is an eye irritant and harmful to plants, is also formed primarily through atmospheric transformation. The rate of atmospheric transformation of these secondary emissions depends on weather conditions.

While considerable field data exist for emissions of CO, VOCs and NO_x, limited test data exist for pollutants like acetaldehyde and PAN. A California study on the air quality effects of ethanol concluded that acetaldehyde and PAN concentrations increase only slightly. The Royal Society of Canada found that the risks associated with increased aldehyde emissions from ethanol-blended fuels are negligible, because emissions are low relative to other sources and they can be efficiently removed by a vehicle's catalytic converter.

There have been fewer studies on the impact on pollution levels of using higher ethanol blends, such as the E85 used in flexible-fuel vehicles (FFVs), but the ones available suggest that the emission impacts are similar. In Ohio, operating data were collected from 10 FFVs and three gasoline control vehicles operating in the state fleet (NREL, 1998). All were 1996 model year Ford Tauruses. As shown in Table 5.2, emissions of regulated pollutants were similar for E85 vehicles operating on E85 and the same vehicles operating on reformulated gasoline (RFG), and were similar to emissions from standard vehicles operating on RFG. Hydrocarbons (including evaporative emissions) were somewhat higher on E85, and NO_x was somewhat lower, but all vehicle-fuel combinations were well below the EPA Tier I emissions standard for each of the four measured pollutants. In the past, FFV and standard gasoline Taurus engines have generally produced very similar NO_x emissions levels. As expected, acetaldehyde (and to a lesser extent formaldehyde) emissions were higher from the E85 fuel. Since these are essentially uncontrolled emissions, in the future aldehyde emissions from all vehicle and fuel types could be much lower if appropriate emissions controls were applied.

Table 5.2

Flexible-fuel Vehicles (E85) and Standard Gasoline Vehicles (RFG): Emissions Comparison from Ohio Study
(grams per kilometre except fuel economy)

Emissions	Flex-fuel (E85) vehicle		Standard gasoline vehicle operating on RFG	EPA Tier 1 standard
	Operating on E85	Operating on RFG		
Regulated emissions				
NMHC	0.09	0.06	0.07	0.16
THC	0.12	0.07	0.08	0.25
CO	0.81	0.62	0.87	2.11
NO_x	0.06	0.05	0.14	0.25
Greenhouse gases				
CO_2	242	255	252	n/a
Methane	0.03	0.01	0.01	n/a
Aldehydes				
Formaldehyde	1.4×10^{-3}	0.6×10^{-3}	0.8×10^{-3}	n/a
Acetaldehyde	8.1×10^{-3}	0.2×10^{-3}	0.2×10^{-3}	n/a
Fuel economy				
L/100km (actual)	14.9	11.1	11.0	n/a
L/100km (gasoline-equivalent basis)	11.0			

Notes: non-methane hydrocarbons (NMHC) and total hydrocarbons (THC) include evaporative emissions. CO_2 emissions estimates are for vehicle only (not well-to-wheels). "n/a": not applicable (no standard for this pollutant). Source: NREL (1998).

Emissions from Ethanol-Diesel Blends

Lower pollutant emissions are one of the primary benefits of using ethanol-diesel blends. Compared to conventional (*e.g.* #2) diesel fuel, ethanol blends of 10% to 15%, along with a performance additive, provide significant emissions benefits. The *Ethanol-Diesel Reduced Emissions Fuel Team* (EDREFT), a multi-industry task force, found that the blending of ethanol into diesel reduces tailpipe exhaust emissions (PM, CO and NO_x) relative to conventional diesel-fuelled engines. For high-level (*e.g.* E95) blends in vehicles with modified diesel engines, the ADM and Hennepin studies (referenced above) show mixed results: E95 trucks had much lower emissions of particulates, and somewhat lower emissions of nitrogen oxides, higher average hydrocarbon (HC) and carbon monoxide (CO) emissions. Swedish tests show much lower emissions for all four pollutants, for E95 buses compared to buses running on Euro 3 diesel: a 92% reduction in CO, 80% reduction in HC, 80% reduction in PM and a 28% reduction in NO_x (Lif, 2002). A number of other studies have also been conducted, with mixed results – though all studies have found significant reductions for PM and NO_x.

Emissions from Biodiesel Blends

The physical and chemical properties of biodiesel are similar to those of petroleum diesel. However, biodiesel has a number of advantages, as shown in Table 5.3. These include better lubricity (lower engine friction), virtually no aromatic compounds or sulphur, and a higher cetane number. Both pure biodiesel and biodiesel blends generally exhibit lower emissions of most pollutants than petroleum diesel. Although emissions vary with engine design, vehicle condition and fuel quality, the US EPA (EPA, 2002b) found that, with the exception of NO_x, potential reductions from biodiesel blends are considerable relative to conventional diesel, and increase nearly linearly with increasing blend levels (Figure 5.1). Reductions in toxic emissions are similarly large (NREL, 2000).

Of particular concern to diesel producers are requirements to reduce the sulphur content of diesel fuel to meet various emissions requirements. Reducing the sulphur content also reduces fuel lubricity. Blending biodiesel can help, since it does not contain sulphur and helps improve lubricity. On the other hand, blending only small quantities of biodiesel with conventional diesel does not bring the average sulphur content down appreciably. To reduce

Table 5.3

Biodiesel / Diesel Property Comparison

	Biodiesel	Low-sulphur diesel
Cetane number	51 to 62	44 to 49
Lubricity	+	very low
Biodegradability	+	–
Toxicity	+	–
Oxygen	up to 11%	very low
Aromatics	0	18-22%
Sulphur	0	0-350 ppm[a]
Cloud point	–	+
Flash point	300-400°F	125°F
Effect on natural, butyl rubber	can degrade	no impact

[a] Ultra-low sulphur diesel has less than 50 ppm sulphur and new diesel regulations in most IEA countries will bring this level to less than 10 ppm by 2010.
Sources: IEA (2002), EPA (2002b), NREL (2000).

Figure 5.1

Potential Emissions Reductions from Biodiesel Blends

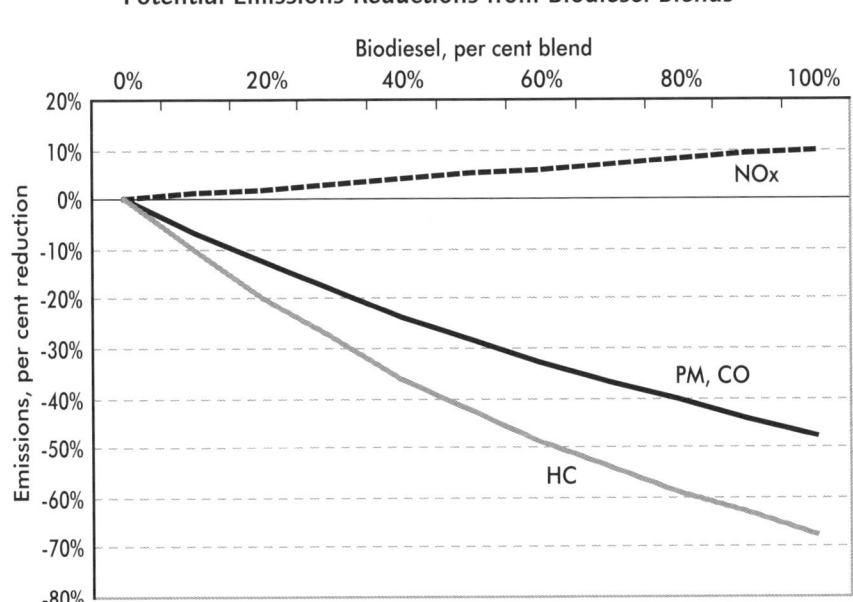

Source: EPA (2002b).

350 ppm sulphur diesel down to 50 ppm, for example, requires a blend of more than 85% biodiesel. At current biodiesel production costs, refiners will likely prefer to cut the sulphur content of conventional diesel at the refinery.

Once the engine is optimised for use with the blend, biodiesel typically raises NO_x emissions by a small amount relative to diesel vehicles. Diesel vehicles in general have high NO_x emissions, so the small increase from biofuels does not appreciably exacerbate this problem. A variety of techniques are being developed to reduce diesel vehicle NO_x emissions to meet emissions standards in the OECD and in other countries (Lloyd and Cackett, 2001). Most of these techniques require the use of very low-sulphur diesel (preferably less than 10 ppm sulphur). B100 or B20 blended with ultra-low-sulphur diesel can be used with many of these techniques.

Other Environmental Impacts: Waste Reduction, Ecosystems, Soils and Rivers

In addition to the "standard" environmental issues surrounding biofuels, such as their contribution to a reduction in GHG and air pollution, there are several other impacts that are often overlooked. These include the impact of biofuels on soils and habitats from growing bioenergy crops, on removing crop and forest residues and using these to produce biofuels, on water quality from bioenergy crop production and biofuels use, and on disposing of various solid wastes (Sims, 2003). The net effect of these factors varies and depends on how the fuels are produced and used, and on the systems and methods applied. In the extreme, such as if a rainforest is replaced by bioenergy crop plantations, the impacts could be strongly negative[4]. In many cases, however, growing bioenergy crops, producing biofuels and using them in vehicles provide net environmental benefits.

There is a clear benefit when biofuels are produced from waste products that would otherwise pose a disposal issue. Both bioethanol and biodiesel can be made from various waste products, *i.e.* crop residues, whey, tallow, cooking oils and municipal wastes. In some cases, the waste product has an economic value, but often society must pay to remove and dispose of this waste. If the

---

4. *This is not the case in Brazil, where most sugar cane plantations (and nearly all new plantations) occur in the south, not far from São Paulo. Most land converted to sugar cane production is grazing land.*

waste product can be used to produce biofuels, then it provides an additional net benefit that may or may not be captured in the price of the biofuel. External costs, such as environmental impacts associated with disposing of waste, are not captured in the market price of biofuels.

Examples of waste products that can be used as biofuels, and can provide social benefits, include:

■ **Municipal waste.** This can be used to produce gas such as methane, which is more commonly used for heat and electricity production, but can be used in natural gas vehicles or converted to liquid fuels. Some municipal wastes have sufficient cellulosic component to convert them into ethanol (via acid hydrolysis), though this could be expensive given the relatively low volumes available and possible requirements for sorting materials.

■ **Crop residues.** While crop residues can contribute to soil fertility, many studies have shown that, depending on the natural soil fertility levels and nature of the specific crop residue, some percentage of crop residue can be removed with little negative effect on soil quality. For example, Lynd *et al.* (2003) estimate that 50% of residues from corn crops could be removed with no detrimental soil impacts. In fact, rather than be ploughed back into the soil, many residues are simply burned in the field (particularly in developing countries). Often they are used for cooking and heating in very low-efficiency stoves or open fires, exacerbating air pollution. Rerouting these wastes to biofuels production can therefore reduce certain environmental and health impacts. In the case of domestic cooking fuel, in order to free up biofuels for other more efficient (and less harmful) purposes, higher-quality fuels generally need to be introduced in homes, such as natural gas or propane.

■ **Forest waste.** Carefully planned removal of plantation and regrowth forest residues can contribute to healthier forests and can reduce the risk of forest fires without disturbing the forest ecosystem. In New Zealand for example, the socio-economic potential of forest residues which will be available for energy purposes as the forest estate matures over the next few years is estimated to be 60-70 petajoules per year – equivalent to a medium-sized natural gas field (Sims, 2003). The residues could be converted to liquid fuels either through techniques like enzymatic hydrolysis or gasification (described in Chapter 2).

- **Waste cooking oil.** The United States produces enough waste greases a year to make 500 million gallons of biodiesel. New York City alone could produce 53 million gallons of biodiesel annually from its waste grease, which is about five times the annual diesel fuel consumption of the city public transit system. The best applied example is in Austria, where quality recycled frying oil was collected from 135 McDonald's restaurants. One thousand tonnes were then transesterified into fatty acid methyl ester (FAME) of standardised quality. Long-term bus trials in the daily routine traffic of the city of Graz have shown full satisfaction when using the McDonald's-based 100% FAME (Mittelbach, 2002; Korbitz, 2002). There are numerous pilot projects exploring this potential resource, including the city of Berkeley, California, which has begun operating recycling trucks (picking up newspapers, bottles and cans) on fuel made from recycled vegetable oil collected from local restaurants.

The net effect of biofuel use on soil quality depends to a large degree on the alternative uses of the land. Converting cropland to energy crops may have minimal impacts (especially if the crops for biofuels are added into crop rotation cycles and soil characteristics are kept in balance). If dedicated energy crops that need little fertiliser or pesticides, such as perennial switchgrass, are mown instead of ploughed, they can enrich soil nutrients and provide ground cover, thus reducing erosion. They may also provide better habitats for birds and other wildlife than annual crops.

Fertilisers and pesticides have an important impact on rivers and other water bodies. While the introduction of low-fertiliser crops, such as grasses and trees, will lower nitrogen release and run-off into nearby water bodies, increasing agricultural activity for grain-based biofuels production may require an increase in the use of fertilisers and pesticides during crop production. Then, nitrogen fertiliser run-off increases the nitrification of nearby water bodies. In such cases, best-practice fertiliser application should be used, with precision farming methods, geographic information systems (GIS), and "little and often" fertiliser application strategies to minimise pollution as well as to lower production costs.

Ethanol-blended gasoline is less harmful to human health and the environment than several other octane-enhancing additives. Lead (Pb) is now being phased out worldwide as its negative impacts on human health are well documented. As lead has been phased out over the past 30 years in IEA countries, methyl tertiary butyl ether (MTBE) has become an important alternative additive for octane enhancement as well as boosting fuel oxygen levels, and is in wide use. It is a somewhat toxic, highly flammable, colourless liquid formed by reacting methanol with isobutylene. However, increasing concern of the potential impacts of MTBE when it mixes with groundwater have caused some countries (and some states in the US) to restrict its use. Ethanol is expected to play an important role in replacing banned MTBE, which will contribute to expected growth in ethanol demand in the US and other countries over the next decade (see Chapter 7).

6 LAND USE AND FEEDSTOCK AVAILABILITY ISSUES

As described in Chapter 2, there are many potential feedstock sources for the production of biofuels, including both crop and non-crop sources (Table 6.1). The potential contribution of each of these feedstocks varies considerably, both inherently and by country and region. It is also related to many factors – technical, economic, and political. In most countries crop-based sources are currently providing far greater supplies of feedstock than non-crop (waste/residue) sources, though at generally higher feedstock prices. The potential supply of dedicated "bioenergy" crops (*e.g.* cellulosic crops), however, is far greater. Further, such crops could supply large quantities of both biofuels and co-produced animal feeds and other products.

Table 6.1

Biofuel Feedstock Sources

Food and energy crops	Biomass wastes / residues
• Divert existing food/feed crops to biofuels (reduction in crop supply for other uses or reduction in oversupply) • Produce more food/feed crops (higher yields, more land, more crop varieties) • Produce dedicated energy crops (cellulosic), partly on new (*e.g.* reserve) land	• Waste oils and tallows • Forest/agricultural waste products • Industrial wastes (including pulp and paper mills, etc.) • Organic municipal solid waste

This chapter begins with a presentation of near to mid-term scenarios of biofuel production in the US and the EU. The scenarios suggest to what extent these countries could displace petroleum transportation fuels with biofuels from locally or regionally produced food/feed crops (*i.e.* grain, sugar and oil-seed crops) using conventional production techniques, as is the norm today. The chapter also examines the potential from other sources, such as dedicated bioenergy crops. Finally, it reviews recent assessments of potential biofuels production worldwide in the very long term, *e.g.* 2050-2100, since most global assessments cover this longer time frame.

Biofuels Potential from Conventional Crop Feedstock in the US and the EU

In North America and the European Union, a variety of initiatives are under way to promote the use of liquid biofuels such as ethanol and biodiesel for transport. As discussed in Chapter 7, both regions are in the process of developing biofuels policies with strong incentives to increase their production and use for transport over the next few years. Market conditions could also drive demand growth. In the US, a number of states appear likely to shift from MTBE to ethanol as a fuel additive which could double US demand for ethanol over the next decade. The renewable fuel targets in Europe, though voluntary, may spur a similar rapid increase in demand.

In both regions, nearly all biofuels are currently produced from starchy and oil-seed crops – mainly grains such as corn and wheat and oil-seeds such as soy and rapeseed. Some cropland is allocated to growing these crops for energy purposes. However, it is unclear how much land can be dedicated to growing these crops for energy purposes, while still meeting other needs (*e.g.* food/feed supply, crop rotational needs, soil supply and quality, and preserving natural habitats).

How much land would be needed for such crops to be used to meet targets for displacement of conventional transportation fuels? The following scenario explores this question for the US and the EU, and estimates the approximate amount of cropland that would be required to produce a given amount of ethanol and biodiesel. The results provide a very rough indication of how much oil could be displaced with domestically produced biofuels, from current crop types, using conventional approaches, over the next 20 years. Later in this chapter several alternative types of feedstock and conversion technologies (such as cellulose-to-ethanol, waste oils and greases to biodiesel, etc.) are discussed that could significantly increase the supply of biofuels beyond the scenario presented here.

Table 6.2 presents crop production and biofuels yields in the US and the EU in 2000, the base year for this analysis. For a variety of reasons, the primary crops for both ethanol production and biodiesel production differ in the US and the EU. Perhaps the most important factor behind the different crop choices is that, in both regions and for both fuels, relatively plentiful crops

Ethanol and Biodiesel Production: Comparison of US and EU in 2000

	Ethanol		Biodiesel		Notes
	US	EU-15	US	EU-15	
Total cropland area of all major crops (million hectares)[a]	133	49	133	49	Includes all major field crops, excludes orchards and grazing land
Crop types currently used for biofuels production and approximate shares	corn 100%	wheat 50% beet 30% barley 20%	soy 100%	rape 70% sunflower 30%	EU production varies by country; see Chapter 1.
Cropland area used for crop types used to produce biofuels (million hectares)	32.2	31.2	30.1	4.1	
Actual cropland area devoted to making crops for biofuels (million hectares) and average yield estimates	2.0	0.1	0.0	0.6	Estimate is based on total production of biofuels
Per cent of cropland area for relevant crops actually used to produce biofuels	6.1%	0.3%	0.2%	14.1%	
Per cent of total cropland area, all crop types	1.5%	0.2%	0.0%	1.2%	
Average crop yield (metric tonnes/hectare)	7.9	7.0	2.5	2.9	Weighted average for current crop types used to produce biofuels
Total crop production (million metric tonnes)	253	160	75	12	
Avg. biofuel yield (litres per metric tonne of crop)	397	400	212	427	Weighted average for current crop types
Avg. biofuel yield (litres per hectare)	3 120	2 790	530	1 230	Weighted average for current crop types
Actual biofuel production (billion litres)	6.2	0.3	0.0	0.7	IEA estimates (note US biodiesel production was 20 million litres, which rounds to 0.0 billion litres)
Biofuel production, gasoline/diesel equivalent (billion energy-equivalent litres)	4.2	0.2	0.0	0.6	Adjusted for lower volumetric energy content of biofuels (ethanol = 0.67 of gasoline, biodiesel = 0.87 of diesel)
Year 2000 consumption of relevant (gasoline/ethanol, diesel/biodiesel) transport fuel (billion litres)	475.0	144.2	173.3	146.0	US and EU data
Biofuel share (volume basis) of relevant road transport fuel	1.3%	0.2%	0.0%	0.5%	
Biofuel share (energy basis) of relevant road transport fuel	0.9%	0.1%	0.0%	0.4%	

[a] Total cropland estimate based on total planted hectares of grain, sugar and oil-seed crops; total available agricultural land is higher and includes fallow and reserve lands, orchards and pastureland.

Sources: United States: USDA, Economic Research Service; except ethanol conversion efficiency from Wang (2001a); European Union: DG Agriculture (crop yields adjusted on the basis of Table 6.3); biofuel production data from F.O. Lichts (2003).

have been used to develop the fuel industry. For example, in the US, where corn production is several times greater than wheat or barley production, corn growers have been the most interested in developing ethanol production as a new product market. In contrast, wheat production is three times higher than corn production in the EU and as a result wheat is the dominant feedstock for the small amounts of ethanol produced (along with sugar beets in France). Similarly, soybeans are the dominant oil-seed crop in the US while rapeseed is in the EU.

The key factors in determining how much land is needed to produce biofuels are crop yields per hectare and biofuels yields per tonne of crop input. There is a fairly wide range in the averages for these factors (Table 6.2). These averages further mask the fact that actual yields, particularly crop yields per hectare, vary considerably by country and over time. Both agricultural yields and conversion yields have been slowly but steadily improving in most regions, and new, large conversion plants may be producing considerably more than the average figures presented here. It appears likely that yields in most regions will continue to improve in the future, at an overall rate of some 1% to 2% per year in terms of litres of biofuels per hectare of land.

Table 6.3 indicates recent average biofuels production rates in each region, based on recent estimates. The table shows that far greater volumes of ethanol than biodiesel can typically be produced from a hectare of cropland, and also that ethanol from sugar crops is much less land-intensive than from grains. Land is not fully fungible between different crop types, however, and all of these crops are only suitable on some types of land or in certain climates. Further, crop-rotation requirements may limit the scope for planting any one crop in any given year.

Given the biofuels yield averages shown in Table 6.3, it is possible to estimate the approximate amount of cropland that was required to produce biofuels in the US and the EU in 2000. The yields are shown for relevant crops – those crops actually used to make biofuels in 2000, and for total cropland, here defined as the total area of cropland planted with field crops (grains, sugars, oil-seeds) during the year. In both the US and EU, a significant share of relevant crops went towards producing biofuels in 2000, including 6% of corn for ethanol in the US and 15% of rapeseed for biodiesel in the EU. But current production of biofuels does not require a substantial percentage of total cropland – the highest share is 1.9% for ethanol in the US.

Table 6.3.

Typical Yields by Region and Crop, circa 2002
(litres per hectare of cropland)

	US	EU	Brazil	India
Ethanol from:				
Maize (corn)	3 100			
Common wheat		2 500		
Sugar beet		5 500		
Sugar cane			6 500	5 300
Biodiesel from:				
Sunflower seed		1 000		
Soybean	500	700		
Barley		1 100		
Rapeseed		1 200		

Sources: Averages estimated by IEA, based on 2000-2002 data and estimates from USDA (2003), EC-DG/Ag (2001, 2002), Cadu (2003), Johnson (2002), Macedo *et al.* (2003), Moreira (2002), Novem/Ecofys (2003).

Although cropland requirements in 2000 were modest, if biofuels production is dramatically expanded in the future, the cropland requirements could become quite significant, and eventually put limits on biofuels production potential. In any case, in order to increase biofuels production, some combination of the following actions must occur:

- Biofuels yields per hectare of land are increased (through improved crop yields and/or improved conversion yields).

- Greater shares of biofuels-appropriate crops are diverted from existing uses to produce biofuels.

- The cropland allocated to biofuels crops is expanded.

- Other types of agricultural land (*e.g.* grazing land) are converted to produce the relevant crops.

The following scenarios look at the potential impacts on crops and cropland if the US or the EU were to expand biofuels production, using the following example targets to illustrate the effects: a 5% displacement of road transport fuel by 2010 and 10% by 2020. In developing these scenarios, a number of factors have been taken into account and assumptions made:

■ Displacing higher percentage shares of transport fuel in the future is made more difficult by the fact that transportation fuel demand is expected to grow – by 32% from 2000 to 2020 in the US and by 28% in the EU (combined increase in gasoline/diesel use, reference case projection from IEA WEO, 2002). However, biofuels yields per hectare of land will also increase. Obviously if future transport fuel demand were lower than the IEA projections, either because of more efficient vehicles or lower vehicle travel, or both, the land required to produce biofuels to displace a certain percentage of transport fuels would be commensurately lowered. But the IEA reference case projection is used here.

■ Although grain and sugar beet production will probably continue to expand in the future, this is mainly because yields are expected to improve. Not much additional land is expected to be devoted to this production under a "base case" scenario (USDA, 2002; EC-DG/Ag, 2002). But crop yields continue to improve, for example with the average for corn in the US increasing from 5.7 to 7.9 metric tonnes per hectare over the last 15 years (about 2% per year). The USDA projects that corn yields per hectare will improve by another 10% over the next ten years, and that soy yields will improve by about 5%. Similar types of improvements are likely to occur for wheat and rapeseed in the EU. In the following scenarios, crop production per hectare for all crops is assumed to improve at 1% per year over the next 20 years. Conversion yields are also assumed to improve, at about 1% per year for ethanol (litres per tonne of feedstock), and at a slower rate (0.3%) for biodiesel, since the process of crushing oil-seeds and converting to methyl ester (biodiesel) is not likely to benefit as much from technological improvements or scale increases (USDA, 2002; IEA, 2000a).

■ There is currently excess production of various types of crops in both the US and the EU. Some of these crops (or the land they are grown on) could be shifted to produce biofuels without requiring a reduction in allocation of crops for other useful purposes. This does not affect the land requirement estimates – it simply means that currently more land is in crop production than needed. This in turn means that requiring a certain amount of land to produce crops for biofuels may take fewer crops away from other purposes than it would otherwise.

■ Only current cropland (and in the EU, set-aside land) is included in these scenarios. The total agricultural area that could be made available for crop production in both regions is much larger; it includes fallow land, conservation reserve land, grazing land, orchards, etc. However, converting such lands to grow the appropriate crops could require large shifts in agricultural practices.

■ It is assumed that vehicles running on low-level blends of ethanol and biodiesel have the same energy efficiency, on average, as those operating on pure gasoline or diesel. As discussed in Chapter 5, research in this area shows a range of potential impacts, and it may be the case that some newer vehicles experience an efficiency boost from low-level ethanol blends (*i.e.* a reduction of energy use per kilometre). However, no firm relationships have emerged from the literature. This assumption is particularly important for low-level blends of ethanol, since ethanol has only two-thirds as much energy per litre as gasoline. For example, an E5 blend (5% ethanol) has 1.7% less energy than 100% gasoline, and (at equal energy efficiency) vehicles need to blend 7.3% ethanol in order to travel as far as they did on the 5% (displaced) gasoline. However, if this blend were to affect overall fuel efficiency by just 1% in either direction, the amount of ethanol required would change substantially. This variation is shown in Table 6.4 for ethanol/gasoline and biodiesel/diesel blends. Thus, while the following scenarios assume no vehicle efficiency impact from blending, it is important to realise that small changes in this factor could have large impacts on total biofuels and land requirements, for a given gasoline or diesel fuel displacement target.

Table 6.4

Biofuels Required to Displace Gasoline or Diesel
(as a function of vehicle efficiency)

Relative energy efficiency of vehicles operating on biofuels blend	Volumetric percentage blend of biofuels required to displace 5% and 10% of gasoline and diesel on an energy basis			
	5% gasoline by ethanol	10% gasoline by ethanol	5% diesel by biodiesel	10% diesel by biodiesel
1% better efficiency	5.9%	13.0%	4.6%	10.3%
Equal efficiency	7.3%	14.2%	5.7%	11.3%
1% worse efficiency	8.6%	15.4%	6.8%	12.3%

Note: The percentages shown are the required volume percentage blends of biofuels in order to displace the target percentage of gasoline or diesel fuel shown at the top of each column. Figures are based on ethanol with 67% as much energy per litre as gasoline, biodiesel with 87% as much energy per litre as petroleum diesel.

The main results of the scenarios for the US and the EU are presented in Figure 6.1. All results and the key assumptions are shown in Table 6.5. Given the foregoing assumptions, for ethanol to displace 5% of motor gasoline in 2010 or 10% in 2020 on an energy basis, some 10% to nearly 60% of "biofuels crops" (*i.e.* crop types likely to be used to produce biofuels) would have to be devoted to biofuels production rather than used for other purposes (such as food or animal feed). The projected crop share in the US is higher than in the EU in part because expected demand for gasoline is forecast to be much higher and in part because proportionately less land is expected to be devoted to growing corn in the US than growing wheat and sugar beet in the EU. In terms of total cropland (*i.e.* land area that is expected to be planted with field crops), a somewhat higher share of land would be needed in the US: about 8% by 2010 to displace 5% of gasoline, and around 14% to displace 10% of gasoline in 2020, versus about 5% of land in 2010 and 8% in 2020 in the EU.

For biodiesel to displace diesel in the percentages specified in these scenarios, much higher crop and land allocations would be necessary than for ethanol/gasoline. Displacing 5% diesel fuel by 2010 would require about 60% of US soy production, and over 100% of projected EU oil-seed (rape and sunflower) production. Thus, in both the US and the EU in 2020, more biodiesel crops would be needed than are expected to be available. Total cropland requirements would again be quite similar in the US and the EU, 13% to 15% in 2010, and some 30% in 2020. Clearly, the amount of cropland that would be needed to displace 10% of diesel fuel is quite large and would require major cropland reallocations towards oil-seed crops used to produce biodiesel. The relatively high land requirements for biodiesel production are due in large part to relatively low yields per hectare compared to ethanol from grain and sugar crops.

So far, these estimates have covered ethanol and biodiesel separately, but if equivalent displacements of gasoline and diesel were sought, the land requirements would become much greater – *i.e.* the sum of the requirements for each. The requirements are shown in the last row of Table 6.5. Based on the assumptions here, it would be quite challenging to meet a 10% displacement of gasoline plus diesel fuel in 2020 in either region, requiring 43% of cropland in the US and 38% in the EU. Therefore it may make sense for countries to focus more on ethanol blending into gasoline rather than biodiesel blending into diesel – at least if land requirement constraints are a concern. Alternatively, as discussed in Chapter 5, ethanol blending into diesel fuel may be worth greater consideration. Another alternative, discussed in Chapter 2, is production of synthetic biodiesel from biomass gasification and Fischer-Tropsch processes, or via hydrothermal upgrading (HTU). These approaches, though currently expensive, yield much higher quantities of diesel fuel per hectare of land than can be achieved via the conventional approach – oil from oil-seed crops converted to fatty acid methyl esters (FAME).

Given the assumptions behind these estimates, they probably represent something close to the "maximum land requirement" case. For example, if vehicles gain an efficiency boost from running on biofuels, especially ethanol, this could significantly reduce land requirements. Land requirements could also be reduced by focusing more on ethanol production than biodiesel production, though there may be refining constraints associated with large displacements

Figure 6.1

Estimated Required Crops and Cropland Needed to Produce Biofuels under 2010/2020 Scenarios

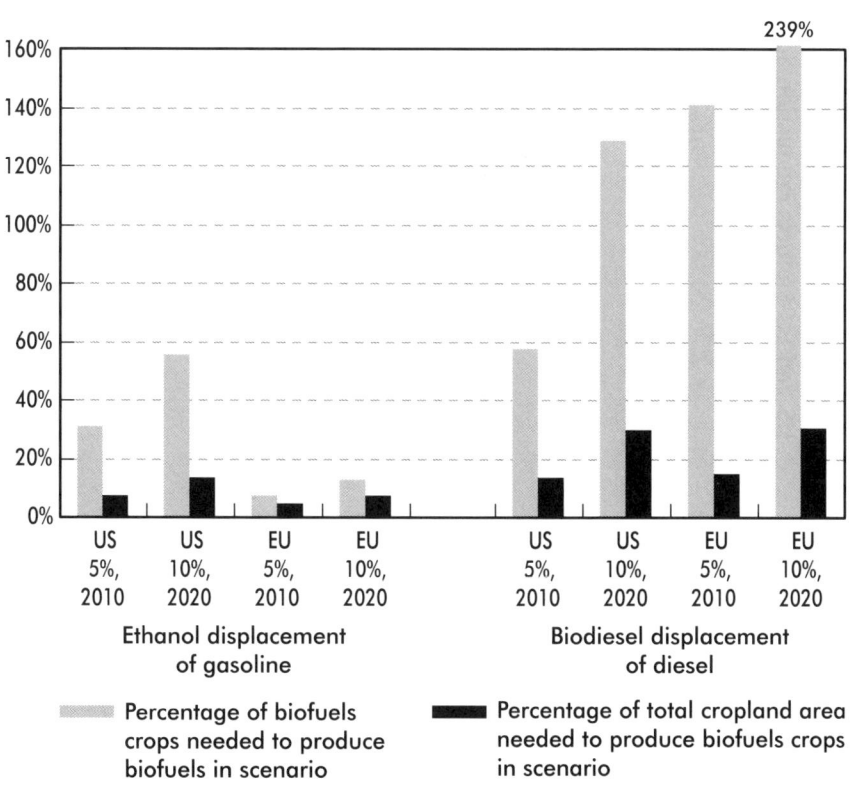

Table 6.5

US and EU Biofuels Production Scenarios for 2010 and 2020

| | 2010 | | | | 2020 | | | |
| | US | | EU | | US | | EU | |
	Ethanol	Biodiesel	Ethanol	Biodiesel	Ethanol	Biodiesel	Ethanol	Biodiesel
Displacement of conventional fuel, per cent (on energy basis)	5.0%	5.0%	5.0%	5.0%	10.0%	10.0%	10.0%	10.0%
Biofuels production under scenario								
Total gasoline/diesel use (billion litres)	535.3	189.6	157.8	178.7	596.0	239.5	164.4	206.3
Required biofuel share of gasoline / diesel pool (volume basis)	7.2%	5.7%	7.2%	5.7%	14.1%	11.3%	14.1%	11.3%
Gasoline / diesel displacement under scenario (billion litres)	26.8	9.5	7.9	8.9	59.6	23.9	16.4	20.6
Required biofuel production under scenario (billion litres)	38.6	10.8	11.4	10.2	84.1	27.1	23.2	23.3
Cropland requirements and availability								
Average biofuels production yields (litres per hectare)	3 800	600	4 800	1 400	4 700	700	5 900	1 600
Cropland area needed for production of biofuels (million hectares)	10	18	2	7	18	40	4	15
Expected cropland area of relevant crops (million hectares)	32	31	30	5	32	31	29	6
Percentage of biofuels crop area needed to produce biofuels	31%	58%	8%	141%	56%	129%	13%	239%
Total cropland area (million hectares)	133	133	49	49	133	133	49	49
Percentage of total cropland area needed to produce biofuels crops for each fuel	8%	13%	5%	15%	14%	30%	8%	30%
Percentage of total cropland area needed to produce crops for both fuels	21%		20%		43%		38%	

Sources: Projections of transport fuel demand from IEA/WEO (2002); US crop production projections from USDA (2002); projections of conversion efficiency are based on 1995-2000 trend and from NREL (as cited in IEA, 2000a). EU land data and crop production projections from EC-DG/Ag (2001, 2002).

of gasoline without similar displacements of diesel fuel. Finally, as discussed in the following section of this chapter, making use of lignocellulosic feedstock could significantly increase the land area and total feedstock available, as well as the overall biofuels yield per hectare of land (for example by utilising crop residues). The extent to which each of these three factors can be exploited in order to maximise potential biofuels production, and develop measures to move in this direction, is worthy of additional study.

In summary, meeting a substantial increase in biofuels demand in the US or the EU over the next 10 to 20 years, using conventional grain, sugar and oil-seed crops, could require a very substantial allocation of cropland – in these scenarios, up to 43% for a 10% displacement of transportation fuel. However, there are other potential sources for biofuels in these and other countries, discussed next.

Ethanol Production Potential from Cellulosic Crops

As discussed in Chapter 2, moving from conventional grain and sugar crops to cellulosic biomass for the production of biofuels opens the door to a much greater variety of potential feedstock sources, including potentially large amounts of waste biomass and more types of land upon which these can be grown. How large is this resource base? Efforts to understand the resource potential for cellulosic feedstock are just beginning, but several recent assessments shed some light on the possibilities.

Two recent US studies, shown in Table 6.6, have developed estimates of cellulosic feedstock supply potential in the US, based on production levels that become cost-effective at various crop prices for different types of biomass feedstock. As the table shows, a significant amount of cellulosic feedstock could be made available for ethanol production at higher prices (though still modest compared to typical prices for grains). Higher production at higher prices reflects both an increasing incentive to farmers to grow dedicated bioenergy crops and a greater area of land where it becomes economically viable to do so[1]. For waste materials, it reflects increasing quantities that become economical to collect and use for biofuels production.

---

1. *In the Walsh* et al. *study, about 50% of land used to grow bioenergy crops is reallocated from traditional cropland, and thus would reduce the amount of other crops. The other 50% would come from idled, pasture and set-aside land. The ratio in the Sheehan study is unclear.*

Walsh *et al.* (2000) provide supply estimates by feedstock source at all four price levels, while Sheehan (2000) provides bioenergy crop estimates only for the higher price levels. Walsh *et al.*'s estimates of cellulosic availability are somewhat higher at all prices and it is unclear from Sheehan's estimates whether some bioenergy crops would become available at a price lower than $50, which could bring the estimates closer together. Both studies find that substantial quantities of crop and forestry wastes could be made available at higher prices, amounting to a greater supply than from dedicated energy crops. Urban and milling wastes would be the cheapest source of cellulosic feedstock, but would contribute the lowest amounts.

These studies do not include some parts of the US for which few data are available. They also make fairly conservative assumptions for feedstock yields per hectare. Thus, the estimates in Table 6.6 do not necessarily represent the full potential of cellulosic feedstock in the US[2]. Conversely, some of the

Table 6.6

Estimated Cellulosic Feedstock Availability by Feedstock Price (million dry tonnes per year)

| | Feedstock cost ($ per dry tonne delivered) | | | |
	$20	$30	$40	$50
Sheehan, 2000				
Crop / forest residues	n/a	37	165	165
Urban / mill wastes	20	21	21	21
Bioenergy crops	n/a	n/a	n/a	150
Total	20	59	186	336
Walsh *et al.*, 2000				
Crop / forest residues	0	25	155	178
Urban / mill wastes	22	71	71	116
Bioenergy crops	0	0	60	171
Total	22	96	286	464

Note: Estimates are cumulative, *i.e.* production level at a price includes the production from the lower price. Prices include costs of feedstock transport to conversion facility. n/a: not available.

---

2. ORNL is currently updating the Walsh et al. *study, and higher estimates of overall potential are expected, but this study was not yet complete as of January 2004.*

dedicated energy crops could take land away from production of other crops, reducing other potential biofuels feedstock supply, but the extent of such competition would be limited by the generally low price for dedicated bioenergy crops. Thus, the potential supply of cellulosic materials appears mainly to complement, and add to, the potential supply from grain/oil/sugar crops.

No studies have been found that focus on the potential supply of cellulosic feedstocks in Europe or other regions, though a similarly large potential is likely to exist anywhere with substantial agricultural, grazing and/or forest land. The studies of world biomass production potential, discussed later in this chapter, generally treat cellulosic feedstocks along with other types in developing their estimates.

Several studies (*e.g.* Kadam, 2000; Novem/ADL, 1999) have developed estimates of potential ethanol production yields from cellulosic feedstock using enzymatic hydrolysis[3]. Conversion efficiencies are estimated to be on the order of 400 litres per dry tonne of feedstock in the post-2010 time frame. By combining either set of feedstock production estimates above with this ethanol yield estimate, the total amount of ethanol that might be produced from the potential US cellulosic feedstock can be estimated for various feedstock prices (Table 6.7). These estimates should be considered long-term potentials, since it will likely be in the post-2010 time frame before a significant amount of cellulosic ethanol conversion capacity will be built.

Using either the Walsh *et al.* or Sheehan feedstock estimates, at a price of around $30 per tonne of feedstock, sufficient feedstock would become available to displace 4% to 6% of US gasoline demand in 2020. At feedstock prices of $50 per tonne, from 23% to 31% of gasoline demand could be displaced. Based on these feedstock data, there appears to be sufficient land available to displace a substantial share of US gasoline demand in the future with cellulosic-derived ethanol.

As mentioned in Chapter 2, the conversion of biomass into liquid using gasification techniques and Fischer-Tropsch synthesis, via hydrothermal upgrading (HTU "biocrude" production) or various other approaches is also under research, particularly in the EU. High yields of biodiesel fuel could be

---

3. *This conversion process is discussed in Chapter 2.*

Table 6.7

Post-2010 US Ethanol Production Potential from Dedicated Energy Crops (Cellulosic)

	$20	$30	$40	$50
Feedstock price and ethanol cost				
Assumed ethanol conversion plant efficiency (litres per tonne of cellulosic feedstock)	400	400	400	400
Feedstock cost per litre of ethanol (US$ per dry tonne delivered)	$0.05	$0.08	$0.10	$0.13
Final ethanol cost ($ per gasoline-equivalent litre – based on Table 4.5, post-2010 NREL estimates)	$0.39	$0.44	$0.47	$0.51
Potential ethanol production by feedstock price				
Using Sheehan feedstock estimates (billion litres)	8.0	23.5	74.4	134.4
Using Walsh et al. feedstock estimates (billion litres)	8.7	38.3	114.4	185.8
Per cent of US motor gasoline consumption, 2010, range[a]	1%-2%	4%-7%	14%-21%	25%-35%
Per cent of US motor gasoline consumption, 2020, range[a]	1%-1%	4%-6%	12%-19%	23%-31%

[a] The range in per cent estimates reflects the two different studies' feedstock availability estimates.
Sources: Plant efficiency and non-feedstock cost based on NREL estimates for 2010 from Table 4.5, as reported by IEA (2000a). US gasoline consumption projection from IEA/WEO (2002).

achieved using these approaches, much higher than biodiesel from oil-seed crops with conversion to FAME. Thus the potential for displacement of petroleum diesel fuel could also be much higher if lignocellulosic feedstocks with advanced conversion processes are considered. Cost reduction is a key issue for these processes.

Another factor that could increase cellulose-to-ethanol yields is improvements in crop yields per hectare planted. While grain crops such as corn and wheat have been improved over hundreds of years, with increases in average yields of several-fold over the past 50 years, very little attention has as yet been paid to potential dedicated energy crops such as poplars and switchgrass. For example, Lynd et al. (2003) estimate that grasses harvested prior to seed production have features that may reasonably be expected to eventually result in a doubling of productivity (in terms of tonnes per hectare) relative to today. This may be possible with or even without the genetic engineering possibilities discussed in Chapter 2. Since switchgrass currently yields about the same total biomass per hectare as corn, this could mean, eventually, much higher relative yields from switchgrass than corn or other grains, if such improvements can be realised. These would also represent much greater improvements than are assumed in the studies above by Walsh *et al.* and Sheehan.

Overall, it appears that if a strong push were made towards development of cellulose-to-ethanol production and other pathways such as lignocellulose-to-diesel, both in terms of feedstock development and conversion technology, the amount of biofuels that could be produced in IEA countries and around the world will eventually be much greater than otherwise. But more work is needed to better understand this potential and how a transition from grain-based and oil-seed-based biofuels to cellulosic biofuels can be encouraged and managed.

Other Potential Sources of Biofuels

Waste oils, greases and fats are low-cost biodiesel feedstock whose availability is not affected by land use policies. A number of studies conducted in the EU over the last several years suggest that the supply of readily collectible waste cooking oil exceeds one million tonnes (Rice *et al.*, 1997). This would be enough to produce about one billion litres of biodiesel, more than was produced (from crops) in 2000, although still a small fraction of diesel use in Europe. Nevertheless, compared to producing biodiesel from crops, the cost of producing biodiesel from waste products is lower.

Currently, the only substantial market for collected oils is the animal feed industry (*e.g.* the UK) or the cement Industry (*e.g.* France). However, tightening controls on animal feed quality may eventually put an end to this usage, and thereby eliminate most of the competition biodiesel producers might face for the supplies. Uncollected (primarily household) waste oils are likely being dumped into sewage systems or landfill sites, although this is illegal in many jurisdictions where waste oil is a listed waste substance.

As mentioned in Chapter 5, one of the few countries with practical experience producing biodiesel from waste oil is Austria, where a total of one million tonnes of recycled frying oil has been collected from 135 McDonald's restaurants. In the United States enough suitable waste grease is produced each year to make as much as 500 million gallons of biodiesel. The City of New York generates enough waste grease from restaurants and other sources to produce 53 million gallons of biodiesel annually, about five times the annual diesel fuel consumption of the city's public transit system (Wiltsee, 1998).

Although, as discussed above, large quantities of lignocellulosic feedstock materials could be made available from crop and forestry wastes, the need to find new ways to dispose of other types of waste could match well with the need to find low-cost feedstock. Landfills worldwide are potential sources of cellulosic materials. Many landfills are close to capacity, even as wastes continue to increase. Likewise, municipalities dispose of tonnes of paper and yard wastes. Other co-mingled wastes amenable to ethanol (or, in some cases, biogas) production could include septic tank wastes, wastewater treatment plant sludge (so called biosolids), feed lot wastes, manure, agricultural wastes, chaff, rice hulls, spent grains from beer production, landscaping wastes, food processing and production wastes (GSI, 2000). However, yeasts producing ethanol are sensitive to the quality and consistency of the feedstock, and specially-designed yeasts and related processes for feedstock purification are being developed in order to increase conversion efficiencies for these types of feedstock.

Overall, waste oils, greases, and lignocellulosic materials represent a very large potential biofuels' feedstock base around the world, and there are many opportunities to obtain these materials cheaply or for free. In general, production scales with waste materials are likely to be smaller than with crop-based feedstocks and, as shown in Chapter 4, the economics of small-scale production of biofuels are generally not as good as for larger-scale production. However, if feedstocks are cheap or free, then the economics improve significantly.

Biofuels Production Potential Worldwide

Two key benefits of biofuels for transport are global in nature: oil savings and greenhouse gas emissions reduction. In both cases, reductions occurring in one country can provide global benefits. With oil savings, a reduction in global demand for petroleum could lower world oil prices and improve security of supply. Greenhouse gases have roughly the same impact on the global climate wherever they are emitted.

This section looks at the potential for growing biofuels feedstock in various regions. This is a fairly new area of research, and many aspects are still uncertain. But enough work has been done to begin to understand the potential global role biofuels could play in the future. Biofuels production potential can be compared to projections of transport fuel demand, to determine the share of biofuels in total transport fuels, and how this may vary by region. Regions where the potential to produce biofuels is high relative to expected transport fuel demand may be interested in exporting biofuels to regions in the opposite situation.

Reliable information on cropland and conversion efficiencies is needed to estimate the potential for global biofuels production. Optimally, one would need to know on what type of land various feedstocks can be grown and how much of that land is available, after taking into account various other required uses for that land. It would also be useful to know the extent to which these feedstocks could be dedicated to biofuels production (as opposed to food, clothing and other materials production, and production of other types of energy such as electricity) and the efficiency of biofuels production per unit land area (taking into account crop production efficiency and biofuels conversion efficiency). One would also need projections for some of these factors. This information is not available for all regions, land types and feedstock types. Where information is missing, estimates have been made.

Several recent estimates of global bioenergy potential are presented in Table 6.8. Most of them are long-term (*e.g.* 2050-2100). Most do not estimate the economic potential, nor do they indicate how the expected bioenergy potential will be attained. The studies provide estimates for biomass energy, not for liquid biofuels *per se*. The potential for liquid fuels in Table 6.8 was calculated using a conversion efficiency factor of 35%. Most biomass for energy purposes is not likely to be used to produce liquid fuels; but rather to produce heat and electricity (and, increasingly, co-production). Nonetheless, the studies provide an indication of the upper limit for global liquid biofuels production over the long term.

The studies differ in many respects, such as in time frame, in the type of estimate (technical or economic) and in the types of biomass feedstock considered. Although not shown in Table 6.8, the studies also make different assumptions regarding expected food requirements, land availability, crop production yields and ethanol conversion efficiency.

Table 6.8 shows a wide range of estimates. On the basis of the most optimistic of the studies, up to 450 exajoules per year of liquid biofuels production is

Table 6.8

Estimates of Long-term World Biomass and Liquid Biofuels Production Potential

Study	Publication date	Time frame of estimates (and low / high for ranges)	Type of estimates (technical or economic potential, feedstock types included)	Raw biomass energy potential (exajoules per year)			Liquid biofuels energy potential after conversion (exajoules per year)[a]	Notes
				Crops (grains, sugars, cellulose)	Biomass wastes (agricultural, forest, other)	Total		
IPCC Third Assessment Report: Mitigation	2001	2050	technical	440	n/a	440	154	Declines due to increasing food requirements
		2100	technical	310	n/a	310	109	
Fischer and Schrattenholzer (IIASA)	2001	2050, low	technical	240	130	370	130	Economic estimate for 2050 assumes continued technology improvements, cost reductions to ethanol
		2050, high	technical	320	130	450	158	
		2050	economic	a/nr	a/nr	150	53	
Yamamoto et al.	2001	2050	"practical" (lower than technical)	110	72	182	64	Assumes declining land availability due to population pressure
		2100		22	114	136	48	
Moreira	2002	2100	technical (crop wastes included in total estimate)	1 301	n/a	1 301	455	Emphasises high efficiencies from co-production of liquid biofuels and electricity
Lightfoot and Greene	2002	2100	technical (just energy crops)	268	n/a	268	94	Looks only at dedicated energy crops, not food crops
Hoogwijk et al.	2003	2050, low	technical	0	33	33	12	Wide range of input assumptions used
		2050, high	technical	1 054	76	1 130	396	

a IEA estimates based on converting the biomass energy estimate in a particular study to liquid fuels at a 35% energy conversion rate. This is similar to the rate assumed by Moreira, Lightfoot and Greene and others when cogenerating with electricity. A slight improvement is assumed for 2050. None of the liquid biofuels potential estimates account for the possibility that some biomass may be used for traditional purposes, which could require up to 50 exajoules.

Note: a/nr: assessed but not reported; n/a: not assessed.

Sources: As indicated in table. Full citations are in references.

feasible, if all biomass available for energy production were used to produce liquid fuels. This is seven times more than the 60 exajoules per year currently used for road transport worldwide. However, as suggested by the wide range of estimates, the practical potential may be much lower. For example, much of the available bioenergy feedstock will probably not be used to produce liquid biofuels. Currently about 42 exajoules per year of biomass energy is used, at very low efficiencies, for household heating and cooking in developing countries (IEA/WEO, 2002). In the *WEO 2002*, the amount of traditional biomass consumed in the residential sector in developing countries is projected to be slightly higher in 2030 than today.

Considerable amounts of biomass may also be used for power generation. Moreira estimates that new, efficient ethanol/electricity plants in Brazil operating on sugar cane and cellulose (from bagasse) can generate 0.31 energy units of ethanol and 0.23 energy units of electricity for each energy unit of input biomass (a net conversion efficiency of 54%). Assuming efficiency improves slightly by 2050, so that the energy conversion to liquid biofuels reaches 35% of the energy into the system, Moreira's biomass potential estimate of 1 300 exajoules would yield about 450 exajoules of ethanol. Studies with lower estimates of total potential, *e.g.* 200 to 400 exajoules of biomass, would yield about 70 to 140 exajoules of ethanol. Production of ethanol could also be lower if biomass were used for other purposes such as fibre production, though the development of biorefineries (as discussed in Chapter 2) could allow co-production of numerous products at high overall efficiency. In any case, the estimates in Table 6.8 of the technical potential for ethanol production are very general and are subject to a number of caveats.

The economic potential for biomass production for energy use is likely to be much lower than the technical potential, and is a function of its cost for electricity and biofuels production, relative to the cost of competing technologies (for biofuels, mainly petroleum cost). One study, that by Fischer and Schtrattenholzer, estimates both technical and economic biomass potential. As shown in Table 6.8, their economic estimate for 2050 is about 150 exajoules, less than half of their lower technical estimate. At a 35% conversion efficiency to biofuels (ethanol), this would imply a maximum economic production potential of about 50 exajoules worldwide in 2050. After accounting for other uses of this biomass, the percentage of transport fuels that could be displaced would be fairly low, though still significant.

Since the studies generally take a long-term view of biofuels potential, they do not provide much insight into how much biofuels could be produced over the next 20 years. At least one study, by Johnson (2002), focuses on global economic potential in this shorter horizon. Johnson estimates the potential for increases in global sugar cane-based ethanol production in the next 20 years. Though sugar cane is just one of many types of biofuels feedstock, as discussed in Chapter 4 it may provide the lowest-cost source of ethanol, at least until full development of cellulosic conversion processes occurs. Thus sugar cane is a logical focus for a near-term assessment of economically viable production potential. Johnson's projections take into account likely cane feedstock production levels, competing uses for biomass (primarily refined sugar) and economic viability[4].

Johnson assumes considerable improvements in cane-to-ethanol yields out to 2020, as production develops around the world and is optimised along the lines of trends in Brazil. Several scenarios are then developed that assume different allocations of cane to sugar, molasses, and ethanol production (Figure 6.2). The "E4" scenario assumes the greatest allocation of cane-to-ethanol production. In this scenario, about 6 exajoules (240 billion litres) of low-cost ethanol could be produced globally, with the largest production in Brazil and India.

Johnson then compares the regional projections for sugar cane ethanol production in the E4 scenario with projected regional fuel demand (Table 6.9)[5], and produces a global transport "balance" (Table 6.10). He assumes 10% ethanol blending for gasoline, and 3% blending for diesel in each region. Under these blending assumptions, some regions, notably India and Brazil, would produce much more ethanol than they would demand. Other regions, notably North America and Europe, would produce far less than they would need.

The scenario indicates that it may be possible to meet a global 10% gasoline and 3% diesel blending target by 2020 using just ethanol from sugar cane. This is an important result because sugar cane is a relatively low-cost biofuel source (as discussed in Chapter 4). There is considerable uncertainty, however, surrounding the level of investment needed and the potential impacts on other markets (especially sugar).

---

4. *Sugar cane ethanol is now close to being cost-competitive with petroleum fuels in Brazil, though perhaps not yet in some other cane-producing countries.*
5. *IEA data adjusted by UN population data to account for differences in regional definitions.*

Figure 6.2

Cane Ethanol Production, 2020, Different Scenarios
(billion litres)

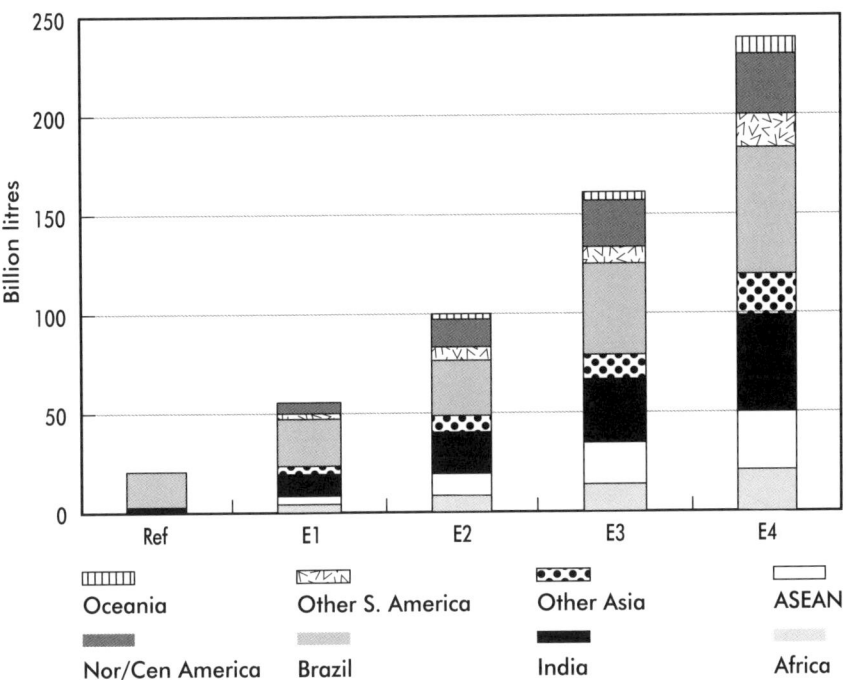

Note: Scenarios E1 through E4 represent increasing allocations of world sugar cane crop and molasses to ethanol rather than sugar production. ASEAN: Association of South-East Asian Nations.
Source: Johnson (2002).

Another noteworthy result of this scenario is how much excess ethanol is produced in India and Brazil. Under the assumptions here, these two countries would easily be able to meet domestic 10% blending requirements and also export substantial amounts of ethanol to other regions, at least through 2020.

The Johnson study analyses the potential for sugar cane to ethanol production. Supplies from other feedstocks could augment the overall supply picture considerably – though probably not at a cost nearly as low as for cane ethanol. In any case, much more analysis is needed on biofuels potential from all types of feedstock, in order to better understand the production potential worldwide, at different cost levels, and how this compares to projected transport fuel demand.

Table 6.9

Current and Projected Gasoline and Diesel Consumption
(billion litres)

Region	Gasoline		Diesel	
	2000	2020	2000	2020
Africa	30	65	34	65
ASEAN	30	63	60	111
India	8	22	43	100
Other Asia	186	397	253	469
Brazil	24	50	3	61
Other South America	30	56	34	56
North and Central America	561	778	242	293
Oceania	22	32	16	21
Europe (including Russia)	242	386	333	439
WORLD	1 132	1 829	1 050	1 614

Source: Johnson (2002), based on IEA and UN projections.

Table 6.10

Cane Ethanol Blending: Supply and Demand in 2020
(billion litres)

Region	Demand 10% gasoline + 3% diesel	Supply (E4 scenario)	Balance
Africa	9	22	13
ASEAN	10	29	19
India	6	49	43
Other Asia	56	23	-33
Brazil	7	62	55
Other South America	8	17	9
North and Central America	88	31	-57
Oceania	4	7	3
Europe / Russia	52	0	-52
WORLD	239	239	0

Source: Johnson (2002).

This study also suggests that there may be a mismatch between where biofuels will be most cost-effectively produced and where they will be consumed. This disparity in turn suggests a need for global trade in biofuels. More work is also needed on this question, to determine the potential benefits – and barriers – to widespread global trade. This is discussed further in Chapter 8.

7 RECENT BIOFUELS POLICIES AND FUTURE DIRECTIONS

As outlined in previous chapters, biofuels feature a number of characteristics suitable for achieving energy, environmental, agricultural and trade policies. As a result, biofuels are emerging as a popular step towards a more sustainable transportation energy sector. In recent years, OECD countries have increasingly advanced policies to support their development – from basic research and development to fuel use mandates – with several countries around the world adopting policies that would require, or strongly encourage, dramatic increases in the production and use of biofuels over the next five to ten years.

As described In Chapter 6, development of international markets for biofuels could markedly change their outlook, and allow OECD countries access to relatively low-cost biofuels produced in non-OECD countries. But first, countries must develop their own industries and infrastructure, and saturate their own potential markets, before international trade is likely to become important.

This chapter reviews recent experiences and current policies in various countries around the world that are leading biofuels producers – or have interest in becoming so. IEA countries are covered, followed by a number of non-IEA countries where interesting developments have recently occurred.

IEA Countries[1]

Canada

In Canada, ethanol first emerged as a blend with gasoline in Manitoba in the 1980s. Today, annual ethanol production is approximately 300 million litres per year and is offered at approximately 1 000 locations in the four western provinces, Ontario and Quebec. The federal government hopes to see an increase in ethanol production by 750 million litres per year, and a number of

---

1. *The EU is also covered in this section.*

major initiatives are under way to boost production significantly over the next few years – potentially with 35% of gasoline containing 10% ethanol by 2010 (Canada, 2003).

The federal government also recently allocated C$ 100 million (US$ 74 million) in its Climate Change Plan to encourage construction of new ethanol plants and development of cellulose-based ethanol. The National Biomass Ethanol Program (NBEP) has C$ 140 million to encourage firms to invest in the Canadian ethanol industry, partially as compensation for a planned reduction or elimination of the ten cents/litre excise tax exemption on fuel ethanol. Participating ethanol producers will be able to draw upon a contingent line of credit if the reduction in the tax exemption impedes their ability to meet scheduled long-term debt servicing commitments (Canada MoA, 2003).

Gasoline quality is mostly regulated by the provincial governments. The Lower Fraser Valley in British Columbia and the southern Ontario region have introduced mandatory vehicle emissions testing. Sales of ethanol-gasoline blends in these regions have increased, as public awareness of the benefits of oxygenated gasoline in passing emissions tests has risen.

The tax system at both the federal and provincial levels has been evolving to support efforts to mitigate climate change and to promote renewable energy and energy conservation. The tax treatment of alternative transportation fuels such as ethanol, propane and natural gas is now favourable nationally and in many provinces, as shown in Table 7.1. In addition to tax breaks, many provinces actually provide tax incentives (subsidies) on ethanol production. This is also shown in the table.

United States

The rise in biofuels use in the US can be traced to the early 1990s (though the first legislation promoting ethanol production and use as a motor fuel was passed in the 1970s). The 1990 Clean Air Act Amendments, and its oxygenated fuels programme, established a requirement that gasoline sold in "carbon monoxide (CO) non-attainment areas" must contain 2.7% oxygen[2]. The reformulated gasoline (RFG) programme requires cleaner-burning reformulated gasoline (requiring 2% oxygen) to be sold in the nine worst ozone non-

---

2. *Ethanol is increasingly used as a substitute for MTBE, and of the 14 cities still in the programme, 10 use only ethanol-oxygenated fuels (EPA, 2002a).*

Table 7.1

Transportation Fuel Tax Rates in Canada (C$ cents per litre)

	Gasoline	Diesel	Ethanol		Ethanol tax incentives
			E85	E10	Per litre ethanol
Federal taxes					
Excise tax	10.0	4.0	1.5	9.0	–
General services tax (GST)	7%	7%	7%	7%	–
Provincial tax					
Newfoundland	16.5	16.5	16.5	16.5	10.0
Prince Edward Island	13.0	13.5	–	–	10.0
Nova Scotia	13.5	15.4	13.5	13.5	10.0
New Brunswick[a]	10.7	13.7	–	–	–
Quebec	15.2	16.2	15.2	15.2	10.0[c]
Ontario	14.7	14.3	2.2	13.2	24.7
Manitoba[b]	11.5	10.9	9.0	9.0	35.0
Saskatchewan	15.0	15.0	2.25	13.5	25.0
Alberta	9.0	9.0	1.4	8.1	19.0
British Columbia	11.0	11.5	0.0	11.0	10.0[d]
Yukon	6.2	7.2	6.2	6.2	10.0
Northwest Territory	10.7	9.1	10.7	10.7	10.0

[a] Alcohol blends are not legal in New Brunswick. [b] For ethanol produced in Manitoba, minimum 10% blend. [c] Incentive is 34.96 cents/litre for ethanol produced in Quebec. [d] 21.0 cents/litre for fuel produced in British Columbia.
Note: Rates are in Canadian currency (C$ 1.0 = US$ 0.74).

attainment areas. About 40 other cities have voluntarily adopted the RFG programme. In addition, the 1992 Energy Policy Act (EPACT) encouraged the use of "alternative fuels". The US federal fleet of vehicles, state fleets and the fleets of alternative fuel providers were required to operate a percentage of their vehicles on alternative fuels. The Clean Cities Program, a voluntary measure under the act, works to create local markets for alternative fuel vehicles. It has worked with cities to develop fleets running on low-blend ethanol and E85, primarily in Midwestern cities close to ethanol production plants.

As a result of these programmes, and promotion of ethanol use by the Environmental Protection Agency (EPA), average ethanol consumption rose by about 2.5% per year during the 1990s. More recently, EPA's requirement that MTBE be phased out in several states, notably in California beginning in 2004, appears likely to lead to an important new driver for ethanol demand, to replace banned MTBE as an oxygenate[3].

During 2003, a US energy bill was passed by both houses of Congress that would considerably increase the support available for domestic ethanol production. However, the combined (House-Senate) version of the bill did not pass on final vote, and it appears that Congress will return to the issue again in 2004.

Taxation of motor fuels in the United States is applied both by the federal government and by state governments. For ethanol there is a federal tax credit of 5.2 cents per gallon of 10% ethanol blended gasoline, yielding effective tax credit of 52 cents per gallon of ethanol, or 14.3 cents per litre. This credit applies to gasoline blends of 10%, 7.7% and 5.7% ethanol (these lower concentrations correspond to 2.7% and 2.0% weight oxygen, required by the 1990 Clean Air Act Amendments mentioned above).

Some states have partial ethanol tax exemptions, particularly in ethanol-producing areas. For example, as of 2002, Idaho had a $0.025 credit for a 10% gasoline-ethanol blend ($0.25 per gallon, or 6.5 cents per litre). Some states also discount sales tax on ethanol, and some provide direct support to ethanol producers.

European Union

The latest policy dealing with biofuels in the EU is contained in two new EU directives, adopted in 2003. One seeks to have biofuels, natural gas, hydrogen and other alternative fuels provide up to 20% of automotive fuel by 2020[4]. National "indicative targets" are now to be set to ensure that 2% of total transport fuel consumption (by energy content) is derived from biofuels by 2005 and 5.75% by 2010. Member States are now developing biofuel strategies to meet these targets.

The directive notes that if 10% of current agricultural land were dedicated to biofuel crops, 8% of current gasoline and diesel consumption could be

---

3. *This should lead to ethanol demand of one billion gallons (3.8 billion litres) per year by the end of the first year of the MTBE phase-out (Schremp, 2002).*
4. *Directive 2003/30/EC.*

replaced with biofuels rather than the current 0.5%. The European Commission noted that expanding ethanol crop production on set-aside land is difficult due to budget constraints from current agricultural subsidies and to the Blair House (trade) Agreement with the US which limits subsidies to rapeseed, soybean and sunflower crops[5].

The second EU directive, adopted in October 2003, addresses the tax treatment of biofuels (within the overall context of energy products taxation)[6]. As a key policy tool for supporting the uptake of biofuels, the EU proposes adjusting fuel excise duties to allow favourable tax deductions for biofuels. Member country rates in 2003, and the EU minimum rates for several transport fuels as from 1 January 2004, are shown in Table 7.2.

Table 7.2

EU Rates of Excise Duty by Fuel, 2003 (euros per 1 000 litres)

	Gasoline (unleaded)	Diesel
European Union: minimum rates as of 2004	359	302
Member country rates, 2003		
Austria	407	282
Belgium	499	290
Denmark	539	406
Finland	597	346
France	586	390
Germany	670	486
Greece	316	245
Ireland	401	379
Italy	542	403
Luxembourg	372	268
Netherlands	631	337
Portugal	508	300
Spain	396	294
Sweden	520	410
United Kingdom	871	826

Note: As of 12/03, one euro equalled about 1.25 US dollars.

---

5. *The Blair House Agreement limits EU oil-seed planting for food purposes to 4.9 million hectares, and plantings for non-food purposes on set-aside land are limited to 1 million tonnes soybean-meal equivalent per year.*
6. *Directive 2003/96/EC.*

Under this new directive, fuels blended with biofuels can be exempted from the EU minimum rates, subject to certain caveats such as the exemption being proportionate to blending levels, with raw material cost differentials, and limited to a maximum of six years. In most member countries, excise duties on diesel and unleaded gasoline in 2003 considerably exceeded these minimum rates, although several countries will need to raise their rates to comply with the new law. Also under the new directive, the assorted temporary and *ad hoc* tax exemptions for biofuels granted to several countries can be continued and extended. As of December 2003, a number of countries have announced such extensions. Table 7.3 provides a list of these countries, and the tax reduction for ethanol relative to the excise duty for unleaded gasoline. In many of these cases, similar reductions are provided for other biofuels like biodiesel, though specific data for biodiesel were unavailable.

Table 7.3

Current EU Country Tax Credits for Ethanol

Country	Reduction in fuel excise duty (€/1 000 l)
Finland	300
France	370
Germany	630
Italy	230
Spain	420
Sweden	520
UK	290

Source: F.O. Lichts (2004).

Another relevant area of recent policy-making is agriculture. The EU is in the process of reforming its "Common Agricultural Policy" (CAP) to make it more compatible with WTO rules, among other things. Objectives include removing crop price support regimes and shifting to a system of support for the agricultural sector more on the basis of good environmental and agricultural practice.

The main CAP reform proposal was approved and adopted in June 2003. The basic agreement de-emphasises crop-specific subsidies. For example, it includes plans for the gradual harmonisation of support prices for cereals, so that non-market price incentives for particular crops will be phased out. There is some separate treatment for energy crops. A trial scheme (to be reviewed in 2006) will provide extra aid of € 45 per hectare of land (except set-aside land) used for energy crop production (*i.e.* crops used for biofuel or biomass power), capped at a total expenditure of € 67.5 million, equal to 1.5 million hectares (Defra, 2003). At an average yield of around 4 000 litres per hectare, this is enough land to produce 6 billion litres of ethanol, about half the amount estimated in Chapter 6 that will be needed to displace 5.75% of gasoline in 2010.

Overall, the current direction of European Union agricultural policy indicates that though some incentives for biofuels will continue to derive from agricultural support policy, this share is likely to diminish over time. Instead, the trend is towards providing fiscal incentives through differential fuel tax regimes, and on environmental grounds.

IEA Europe[7]

Finland is a world leader in the utilisation of wood-based bioenergy and biomass combustion technologies. The primary focus has traditionally been on power generation, but interest in liquid biofuels is increasing, and the government recently began a pilot project with an ethanol plant for production of E5, to produce about 12 million litres per year.

France is a major producer of biofuels, producing both ethanol and biodiesel in large quantities. While production has been stable in recent years, France will probably respond quickly to the new EU targets and tax policies, and may extend tax credits and subsidies as the favoured support instrument.

Historically, **Germany** has not strongly promoted fuel ethanol and has made much more use of biodiesel (both pure biodiesel and blended are increasingly available at gasoline stations, and Volkswagen was the first European car manufacturer to extend warranties to cover use of biodiesel). In November 2003, the government proposed changes to the tax law in accordance with the new EU directive, to exempt biofuels 100% from gasoline taxes for a period of six years. The exemption is granted for blends of up to 5% bioethanol, and only for undenatured alcohol (undenatured alcohol faces the higher import duty of € 19.2 per hectolitre, so the tax exemption will provide a stronger incentive for European-sourced alcohol than for imports).

---

7. *The following reflects recent news reports and general trends in each country; specific sources are not cited for most of this discussion.*

Italy produces some biodiesel, and has recently created ethanol incentives of a tax break of 43% for three years, to support ETBE production.

The Netherlands. Dutch plans for biofuel production are mostly still in the planning stage. While biomass energy is a key energy priority, the Ministry for Economic Affairs is currently still investigating the role biofuels could play in the Dutch energy policy strategy.

Portugal. The Portuguese government has recently approved financing of 50% of building costs for a biodiesel plant (approximately € 12.5 million), due on-stream in July 2004.

Spain is the largest producer of fuel ethanol in the EU, with plans to increase ethanol production over the next two years to over 500 million litres and to expand biodiesel plant capacity as well. Both national and regional governments provide subsidies for plant construction and for promoting ethanol use. The main ethanol producer, Abengoa, receives a 100% tax deduction, and the provincial government of Castilla-León, for instance, will offer support for its renewables target of achieving 9-12% renewable energy by 2010.

Sweden uses both high and low blends of ethanol, including 250 million litres of E5, roughly 50 million litres of which is domestically produced and the balance is imported. A new ethanol plant, with 50 million litre per year capacity, is being built in Norrköping in eastern Sweden, south of Stockholm, after some years of delays and uncertainty and the final granting of limited tax incentives (tax excise reductions worth € 132 million for 2003). The ethanol produced (from grain grown on set-aside land) is to be added to the E5 gasoline sold in the Stockholm area.

The **United Kingdom** has a number of initiatives for promoting alternative fuels and "low-carbon vehicles". However, current support for biofuels is limited. There is no significant UK production of biofuels and only limited plans, mainly for small-scale biodiesel production facilities. That said, the UK makes significant use of fiscal incentives for a wide range of clean fuels. Vehicle excise duty is differentiated according to vehicle CO_2 emissions and by fuel type (Table 7.4), and fuel duty is also differentiated, with a significant tax break for biofuels – as of 2003, 41 eurocents per litre, compared to 78 eurocents for unleaded gasoline.

In **Norway**, the use of biofuels is mostly limited to field experiments. The Norwegian government is taking a market-based approach, and there is no

Table 7.4

UK Annual Vehicle Excise Duty for Private Vehicles
(British pounds per year)

Vehicle CO_2 emissions range (g/km)	Diesel vehicles	Gasoline vehicles	Alternative fuel vehicles
Up to 100	75	65	55
101 to 120	85	75	65
121-150	115	105	95
151-165	135	125	115
166-185	155	145	135
Over 185	165	160	155

Source: UK DVLA (2004).

national goal regarding future use of renewable fuels in the transport sector. However, the government is willing to subsidise research projects in the area of renewable fuels as well as the first phase of commercial use of these fuels. Biofuels are exempted from fuel taxes (except VAT). There have been no practical experiences with ethanol as a motor fuel. In the 1990s, fleet tests in Norway were mostly focused on natural gas and electric vehicles.

IEA Asia-Pacific

As part of **Japan's** plan to meet its emissions reduction target under the Kyoto Protocol, Japan is introducing gasoline with 3% ethanol in 2004. The government targets 10% ethanol blends as the standard by 2008. There are ongoing tests regarding ethanol and vehicle engine compatibility. The Ministry of Environment also plans to set up and to subsidise low-concentration blended fuel pumps at gasoline stands in some regions. The ministry has urged the automobile industry to produce models warranted for using gasoline containing 10% ethanol. If Japan eventually adopts an ethanol blending ratio of 10%, its ethanol market is projected to be around 6 billion litres per year.

Japan has no surplus agricultural production and will probably import biofuels, with relations being developed with Brazil and Thailand. Mitsui, a large trading firm, signed an import pact with Brazil in 2001, and estimates that a market of 6 billion litres per year would develop if 10% blending were implemented throughout the country.

Australia is rapidly becoming interested in biofuels for transport for two reasons: it is committed to limit greenhouse gases from its transport sector, and it has an enormous agricultural base from which to draw feedstock. Current commercial production of biofuels, blended into gasoline, is small – about 50 million litres per year (primarily ethanol), or 0.2% of Australia's gasoline demand. At present, Australian ethanol is produced mainly from wheat, but a study recently conducted by Australia's National Party found that an additional 300 million litres of ethanol could be produced from "low-cost sources", such as sugar cane molasses, by 2010.

The main ethanol producer, Manildra Park Petroleum, produces a 20% ethanol-blended gasoline that is sold in 200 gasoline stations in New South Wales. More recently, BP began running a trial ethanol-blending facility at its Bulwer Island refinery near Brisbane with an E10 blend for Queensland's east coast market.

In 2001, the Australian government adopted a pro-ethanol policy, including eliminating the excise tax. There were strong objections to this programme, primarily from oil companies and car manufacturers, over how much blending can be tolerated by vehicles, whether biofuels are a suitable substitute for MTBE, whether subsidies should be considered for the sugar industry, and what would be the impact on the national budget. Much of the controversy stemmed from the technical debate over the compatibility of ethanol mixes with gasoline and conventional vehicles. These are valid concerns which are discussed in Chapter 5. The percentage of ethanol mixed with gasoline varied from state to state, from 24% in ethanol-producing states to zero in others. Within this range, but primarily for blends above 10%, reports of poor operation and even engine damage became widespread during 2001 and 2002. Australia's oil refining industry and car makers have become reluctant to support ethanol, and some companies have opposed the government mandating its use.

In September 2002, the government announced changes to the policy, including setting a 10% limit to blends, and re-instituting an excise tax on ethanol and other biofuels. However, a biofuel domestic production subsidy, equivalent to the excise duty (A$ 0.38, about US$ 0.24, per litre) was implemented concurrently, resulting in an effective import duty at the value of the excise tax. The subsidy programme is due to be reviewed during 2004. In July 2003, the government announced an additional production subsidy for ethanol plants at the rate of A\$ 0.16 (US\$ 0.10) per litre, available until total domestic production capacity reaches 350 million litres or by end 2006, whichever is sooner. The maximum total cost of the subsidy will be A\$ 49.6 million over five years.

New Zealand's Environmental Risk Management Authority authorised, in August 2003, the sale of fuel ethanol derived from sugar, starches and dairy by-products, for blending with gasoline in the range of E1 to E10.

As part of **South Korea's** plan to expand the use of environment-friendly fuels, in 2002 it commenced the sale of diesel containing biofuels from rice bran, waste cooking oil and soybean oil. An LG-Caltex Oil gas station selling biodiesel was opened in June 2002 in a test area. During the testing period which runs until May 2004, cleaning and garbage trucks will use the biodiesel starting from landfill areas. A decision whether to designate biodiesel as an official motor fuel is to be made after the test period. The Ministry of Energy expected that South Korea could save 30 000 barrels of diesel per year, accounting for 0.02% of the country's total diesel consumption, if all the vehicles running in metropolitan landfill areas used biodiesel.

Non-IEA Countries

Eastern Europe

EU-candidate countries (CCs) in Eastern Europe have a large, mostly unexplored potential to produce biofuels. Since the EU is heavily dependent on imported energy resources, especially oil, and is promoting biofuels in road transport, some EU countries are considering ways to tap the potential in Eastern Europe in order to meet their targets under the proposed directive on biofuels.

A 2003 study by the Institute for Prospective Technological Studies found that the potential contribution of the 12 CCs to the EU-27's biofuel consumption would likely be relatively modest – but not insignificant – at around 1% to 3% for bioethanol and 1% to 2% for biodiesel, under various scenarios (IPTS, 2003).

The study also found that biofuel production in the CCs may not be less expensive than in the EU-15. Production costs, excluding taxes and subsidies, per litre of biofuel in the CCs vary significantly: € 0.41 to € 0.75 per litre for biodiesel and € 0.36 to € 0.60 for bioethanol. These figures are similar to the average current production costs of biofuels in the EU-15, discussed in Chapter 4. Thus, the CCs can contribute positively to the biofuel supply of the EU, but they probably will not contribute a massive, inexpensive supply. More specific information for certain Eastern European countries follows.

Poland produces 50 million litres of biofuels, down from a high of 110 million litres in 1997. However, following introduction of the EU 2001 directive on biofuels, the Polish Parliament adopted a strategy for further development of biofuels by the year 2010. Current proposals call for liquid fuels sold in Poland to contain a minimum 4.5% of bioethanol, raised to 5% in 2006. The Council of Ministers also has proposed that eco-components (such as biofuels) would be exempt from excise tax.

The biofuel strategy should stimulate development in Poland's rural areas. In 2001, rapeseed was planted on 560 000 hectares of land. The government estimates that Polish farmers could produce 2.5 million tonnes of biofuel and fodder from 1 million hectares of rapeseed. But the current debate over biofuel regulations in Poland is contentious. The Polish government has faced open resistance regarding its pro-biofuel policy from oil and gas fuel producers, car producers, and even from the Ministry of Finance, as the biofuels excise tax exemption will decrease budget revenues coming from excise and VAT taxes. If the proposed blending targets and excise tax break become law, the Polish market would need 260 000 tonnes of dehydrated alcohol and 400 000 tonnes of rapeseed oil to meet preliminary requirements.

The **Czech Republic** Ministry of Agriculture provides subsidies for the production of biodiesel from rapeseed oil and for bioethanol. Financial support is limited to Kcs 3 000 (about US$ 90) per tonne of methyl ester and Kcz 15 (US$ 0.45) per litre of bioethanol. In 1999, the Ministry of the Environment spent Kcz 66 million to support the production of 23 thousand tonnes of biodiesel fuel. The ministry also spent about Kcs 10 million to support production of 650 thousand litres of bioethanol (UNFCCC, 2003).

The biodiesel programme in the Czech Republic commenced in 1991 and today most filling stations offer biodiesel. Recently, the Czech Republic has had a surplus of cereal crops with limited possibilities for export. This has spurred the evaluation of the use of cereals in industrial processing to produce ethanol and to blend it with gasoline or use it in the chemical industry. Some 600 000 tonnes of cereals are anticipated to be processed for ethanol production in 2005 (AGRI, 2002).

Hungary is also interested in developing a domestic ethanol market, and has already removed excise taxes on ethanol at the pump. MOL, the Hungarian oil and gas company, has expressed interest in increasing the production of ethanol and blending it with gasoline.

In the **Ukraine**, there is a rapidly growing ethanol industry, both for domestic use and for export to other European countries. The Ukraine has 46 ethanol production facilities, owned by the UkrSpirt Conglomerate. In 1999, the Ukrainian Parliament passed a law which allowed a high-octane oxygenate additive to be used, blended at 6% with gasoline. The excise tax applicable to the 6% blended gasoline was 50% lower than the tax on unblended gasoline. As a result, in 1999 and 2000 about 22 million litres per year were produced. This tax incentive has since been discontinued. However, the city of Kiev, with support from the Ministry of Health, has launched a pilot project for the use of 6% blended fuel in public transport. If the results of this pilot project are positive, the Kiev City Administration plans to make 6% blended fuel mandatory for public transportation.

Latin America

The **Brazilian** Alcohol Programme (Proalcool), launched in the 1970s, remains the largest commercial application of biomass for energy production and use in the world. The undertaking involves co-operation between the Brazilian government, farmers, alcohol producers and car manufacturers. It succeeded in demonstrating the technical feasibility of large-scale production of ethanol as a transport fuel, and its use in high-level blends as well as in dedicated ethanol vehicles.

Prompted by the increase in oil prices, Brazil began to produce fuel ethanol from sugar cane in the 1970s. Production increased from 0.6 billion litres in 1975 to 13.7 billion litres in 1997, by far the highest production of fuel ethanol in the world. The task for the first five years was to displace gasoline with E20 to E25 (20% to 25% blends of ethanol with gasoline). This was completed without substantial engine modification in light-duty vehicles.

Vehicles produced for sale in Brazil are generally modified to run optimally on these blend levels. Idle production capacities and the flexibility of existing distilleries were used to shift production from sugar to ethanol.

After the second oil crisis (1978/79), steps were taken to use hydrated, "neat" ethanol (typically 96% ethanol and 4% water). The Brazilian car industry (*e.g.* Volkswagen, Volvo Brazil, etc.) agreed to implement the technical changes necessary for vehicles to safely operate on the neat fuel. The investment required for this phase of the programme was funded through soft loans by the government. Furthermore, tax reductions made the ethanol option highly attractive to consumers. By December 1984, the number of cars running on pure hydrated alcohol reached 1 800 000, or 17% of the country's car fleet (Ribeiro, 2000). By the late 1980s, neat ethanol was used in over a quarter of cars (3-4 million vehicles consuming nearly 10 billion litres per year; the remaining vehicles used blends of 22-26% ethanol (4.3 billion litres) (FURJ, 1998).

The sharp decrease in oil prices in the mid-1980s greatly increased the relative cost of fuel ethanol production and this was coupled with the elimination of government subsidies for new production capacity, and rising costs from the ageing distribution system. The decree establishing Proalcool and related regulation was revoked in 1991. Ethanol supply shortages raised concerns about driving neat-ethanol vehicles and lowered demand for fuel ethanol, particularly for E96. The share of neat-ethanol vehicles fell from almost 100% of new car sales in 1988 to fewer than 1% by the mid-1990s. However, including the new type of flexible-fuel vehicles (that can run on up to 100% ethanol), recent sales have experienced a resurgence: production of neat and flex-fuel vehicles was 56 000 in 2002 and 85 000 in 2003, or about 4% and 7% of the new car market, respectively. The total stock of ethanol cars peaked at 4.4 million in 1994, and while by September 2002 it had dropped to 2.1 million vehicles, it is likely to soon be increasing again (FURJ, 1998; DATAGRO, 2002).

At the same time, demand for fuel ethanol for blending is rising rapidly. Brazilian ethanol demand is on track to increase by another 2.9 billion litres per year, almost 20%, by 2005 (F.O. Lichts, 2003). Currently, gasoline blending with 20% to 25% anhydrous ethanol is mandatory for all motor gasoline sold in Brazil. This rule created the stability necessary to allow the automotive industry to accelerate the widespread introduction of technological innovations: virtually all new cars in Brazil have the capability to safely operate on the 20-25% blend of gasohol (Ferraz and da Motta, 2000).

In 2002, the Brazilian government began reviving the Proalcool programme. The Industrial Production tax was reduced for manufacturers of ethanol-powered cars, as well as subsidies for the purchasers of new ethanol cars. The government also introduced credits for the sugar industry to cover storage costs, in order to guarantee ethanol supplies. At the heart of the government programme is a 10-year deal with Germany. Germany will purchase carbon credits as part of its Kyoto Protocol commitments and, in turn, will help Brazil subsidise taxi drivers and car hire companies by R$ 1 000 (US$ 300) per vehicle on the first 100 000 vehicles sold (*Kingsman News*, 2003).

Ensuring sufficient, secure ethanol supplies, particularly between one sugar cane harvest and the next, is considered crucial to the success of the government's efforts to revive the Proalcool programme and to rebuild consumer confidence in ethanol-powered cars. The government has developed a programme to build up ethanol stocks during harvest periods, funding the supply build-up and paying for this by selling ethanol during draw-down periods. About R$ 500 million has been allocated to this programme since 2001. The government asked the industry to produce an additional 1.5 billion litres of alcohol from the 2003/04 crop to be added to stock; to maintain a maximum alcohol price at 60% of the gasoline price; and to commence the harvest in March to boost available alcohol supplies. In the meantime, a glut of alcohol emerged during winter 2003/04 and alcohol prices plummeted (see Chapter 4). It now appears that there will be little chance of high ethanol prices or supply shortages during 2004.

Despite periodic ethanol shortages, Brazil is increasingly hoping to strengthen the market by looking to increase exports. Brazil's President recently told representatives of the industry that there was the potential for Brazil to double its ethanol output over the next few years in order to accommodate growth in demand from other countries. To this end, representatives from Brazil's sugar/ethanol sector and from 19 sugar cane states met in September 2003 to formulate a plan to promote the opening of a global ethanol market; and the main strategy will be to persuade the government to draw up institutional export plans to countries that use ethanol as an additive in gasoline. Brazil is the world's largest ethanol producer and has the best technology – aspects which combine to generate export opportunities that many other countries do not have. Brazil is currently negotiating with a number of countries, including China, Japan, South Korea, the US and Mexico, that have expressed interest in buying Brazilian ethanol. While the US is one of the nearest and potentially biggest markets, agricultural subsidies and import restrictions frustrate Brazil's export efforts there.

Peru is particularly well-suited to produce sugar cane and could competitively produce ethanol, both for domestic use and for export. However, to realise the country's potential, the Peruvian government will have to adopt a clear policy to stimulate production.

The government's current objective is to eliminate leaded fuels by 2004. Peru, like other countries that have phased out leaded fuel, will need to develop alternative octane enhancers, and ethanol is one such possibility. The government is also seeking to export ethanol to the growing California market by December 2004. To do so, Peru plans to produce up to 25 000 barrels per day of sugar-based ethanol, as part of a $185 million project planned to be on line by late 2004. The project will include construction of several sugar cane distilling facilities and a pipeline to transport the ethanol from distilleries to the Bayovar port some 540 miles (900 km) north of Lima. In preparation for the project, 2 670 acres (1 080 hectares) of sugar cane-for-ethanol feedstock have been planted in the central jungle (BBI, 2003).

The **Costa Rican** government is very interested in biofuel options and has recently announced plans to begin substituting ethanol for MTBE in gasoline. This move reflects a convergence of trade, energy and environmental concerns. The Costa Rican economy has been strongly affected by external market forces, both for petroleum imports and for exports of its basic tropical commodities like coffee, sugar and bananas. From 1980 to 2000, coffee exports doubled; but, as coffee prices almost halved over the same period, the total value of coffee exports remained basically the same and their relative contribution to the country's trade balance declined significantly. During the same time, the demand for gasoline and diesel fuels and the number of new cars grew rapidly, requiring the country to import more and more oil. Such pressures have led the government to explore new fuel options for transportation and electricity production.

The Costa Rican sugar industry has the potential to supply feedstock for the production of ethanol. In 2001, sugar production was 7.1 million bultos (equivalent to 50 kg each) and total exports amounted to 3.3 million bultos – implying that there is considerable potential to produce more ethanol – both for domestic use and for export (Vargas, 2002).

The Costa Rican Government's National Plan for Development (2002-2006) includes a mandate for the substitution of MTBE with ethanol in gasoline. To implement this mandate, authorities have brought together representatives from the Ministries of Agriculture and Environment, and major interest groups. The first initiative of the group will be to quantify the potential for sugar-based ethanol to replace MTBE. Once the market potential is defined, government subsidies and regulations are expected to assist with development.

In the **Caribbean**, under the Caribbean Basin Initiative (CBI) there is no United States tariff on ethanol imported from this region. Elsewhere, there would be a 52 cent-per-gallon (14 cent per litre) import tax on the fuel (DA, 1999). However, the CBI ethanol programme is capped at 7% of the total amount of US ethanol.

Asia

India has a large sugar cane industry. In 2000, it produced about 1.7 billion litres of ethanol (for all purposes), more than was produced in the EU. Ethanol is now produced mainly from sugar cane-base molasses, but there are good prospects for producing it from other sources, such as directly from sugar cane juice and, eventually, cellulosic crops. Until recently, ethanol production was used primarily for non-fuel (industrial, beverage and pharmaceutical) purposes.

During 2002, a number of projects were initiated, involving blending ethanol with gasoline and selling it at retail fuel outlets. As of mid-2003, about 220 retail outlets in eight districts have sold 11 million litres of ethanol for blending. Six more projects have been approved for ethanol-gasoline blending. Biofuels will eventually be provided in over 11 000 retail outlets after full phase-in of the blending programme.

On 1 January 2003, India implemented a new programme to encourage a rise in ethanol production and use for transport. In the first phase, nine Indian states and four union territories began phasing in a 5% ethanol blend in gasoline. The second phase, to be initiated before year-end 2004, will spread the programme nationally. A third phase will then see the blend increased to 10%. There are also plans to blend ethanol with diesel. The government has proposed a National Biofuel Development Board to oversee the plan. In addition, India and Brazil have signed a Memorandum of Understanding related to ethanol sales and technology transfer.

While the plan is a part of India's recent efforts to cut oil imports, improve urban air quality and to promote more climate-friendly fuels, it is also designed to assist and stimulate the domestic sugar industry. The government ensures a price to sugar millers fixed at Rs 15 (about $0.33) for every litre of ethanol they produce, representing a sizeable subsidy over production costs estimated to be as low as Rs 7 ($0.15) per litre. As the programme is implemented, the number of sugar plants opting for ethanol production is likely to increase dramatically. Of the 196 registered sugar co-operatives (which between them had unsold stock of 4.3 million tonnes of sugar valued at Rs 5.3 billion at end of 2002), as of 2003, 25 had already requested and received licences for ethanol production.

China, in order to meet growing demand for gasoline, has selected several provinces to use trial blends of 10% ethanol. China is the third-largest ethanol producer in the world, with annual production of around three billion litres. Corn is the primary feedstock, but distilleries are also experimenting with cassava, sweet potato and sugar cane. The industrial use of corn is set to rise sharply, boosted primarily by increased demand from the ethanol industry.

In recent years, China has been stepping up the expansion of its ethanol industry in major corn-producing regions. By 2004, a pilot plant to produce 600 000 tonnes per year of ethanol will have been completed in Jilin province. In Shandong, Heilongjiang and Inner Mongolia, a number of projects have also been initiated. A full-scale plant designed to produce 300 000 tonnes of fuel ethanol in Nanyang, in Henan province, should be completed by 2004.

Chinese sugar industry executives from the north-eastern state of Heilongjiang have also been to Brazil to observe its production techniques, policy approach and investigate the possibility of importing Brazilian fuel ethanol.

Thailand aims to increase ethanol production in order to reduce its oil import bill and to create new outlets for farm produce. In 2000, the Thai government declared its intention to promote the use of biofuels produced from indigenous crops, such as sugar cane and tapioca. So far, two oil companies have started distributing fuel ethanol blends, but they have had difficulties sourcing sufficient quantities of feedstock. However, as part of Thailand's fuel ethanol programme announced in late 2000, the government plans to stimulate production to 650 million litres per year in the near term. Local and international investors are looking at the possibility of building plants based on sugar cane and tapioca feedstock. The Thai government recently approved the construction of eight new plants and is looking at permitting the construction of a further 12 using cassava, cane molasses and rice husk as feedstock to produce about 1.5 million litres of ethanol daily (*Kingsman News*, 2003).

The government has also introduced a package of tax incentives to stimulate production, including exemptions on machinery imports and an eight-year corporate-tax holiday. The state-run Petroleum Authority of Thailand (PTT) has been asked to co-invest with private ethanol-blending plants, giving an assurance of state support. On the consumer side, the Finance Ministry will impose only a nominal excise tax, expected to be about \$0.02-0.03 per litre on ethanol-based fuels. State agencies have also been requested to use ethanol fuel in their vehicles over the next two years to promote its consumption.

In addition, the Japanese Marubeni Corporation is working with the Thai company Tsukishima Kikai to complete a commercial plant in Thailand in 2005. The plant is efficient at extracting alcohol from sugar cane wastes and tapioca using genetically modified bacteria. It will cost between 2 and 3 billion yen to build and will produce 30 million litres of ethanol a year. The ethanol will be sold in Thailand but could be exported to Japan. The Itochu Corporation is preparing to enter the ethanol business and is planning to build a demonstration plant in Japan to produce ethanol using wood wastes as a raw material.

Malaysia produces about half of the world's palm oil, which in turn is the vegetable oil with the greatest worldwide production. It is also the oil plant with the highest productivity (tonnes of oil per hectare of land). As a result, Malaysia has begun to export palm oil for the purposes of producing biodiesel and to construct biodiesel facilities within the country. Only about 6 000 litres per year of biodiesel are currently produced in Malaysia, but this could rise rapidly with the construction of new plants.

Africa

Africa has the world's highest share of biomass in total energy consumption, mostly firewood, agricultural residues, animal wastes and charcoal. Biomass accounts for as much as two-thirds of total African final energy consumption, compared to about 3% of final energy consumption in OECD countries. Firewood accounts for about 65% of biomass use, and charcoal about 3%. Currently very little biomass in Africa is converted to liquid fuels. However, bagasse (sugar cane, after the sugar is removed) supplies approximately 90% of the energy requirements of the sugar industry throughout Africa and could make an important contribution to the availability of liquid fuels via ethanol production.

The rising cost of gasoline, lead (Pb) phase-out programmes, and the declining cost of producing ethanol and sugar cane have created favourable economic conditions for fuel ethanol production in Africa. In many countries, lead additives are still heavily used in gasoline, and sugar cane production is abundant – creating the opportunity to use ethanol as a viable alternative source of octane. More than enough sugar cane is produced in Africa to replace all the lead used in African gasoline – a level which would require Africa to produce about 20% of the amount of ethanol currently produced in Brazil; and which would require the shift of only a modest share of sugar production to ethanol production. Countries like Zimbabwe, Kenya, Egypt, Zambia, Sudan, Swaziland, and Mauritius could replace lead with ethanol using primarily the sugar by-product, molasses.

The **Republic of South Africa** accounts for approximately 70% of the continent's total ethanol production, although the majority of it is high-purity ethanol destined for industrial and pharmaceutical markets. Production of high-purity ethanol has been growing in recent years, with the total in 2001 reaching 126 000 tonnes, against 97 000 tonnes in 2000. Only small volumes of fuel alcohol are produced now and larger plants will likely be needed by January 2006 to produce enough ethanol to replace lead, as part of the government's programme to phase out leaded fuel.

Ghana plans to begin production of biodiesel from physic nuts in 2004 and expects to save about $240 million on imported diesel. The first phase of a $1.2 million factory that will produce the fuel is near completion at Pomadze in the central region. It will have an initial capacity of 3 600 tonnes (about

4 million litres) but production is expected to expand. The physic nut yields about nine tonnes per hectare and can be harvested from the fifth month after cultivation. It can achieve maturity from the third year after planting.

Ghana, Mali and other African countries have also been considering the production of biodiesel from oils extracted from the common Jatropha plant, which is tolerant of poor soils and low rainfall.

Outlook for Biofuels Production through 2020

Given the recent trends in biofuels production shown in Chapter 1, and having reviewed recent policy activity in many countries around the world, it is possible to offer some projections of where things seem to be heading. Figure 7.1 Illustrates both recent trends (bars and, for world, dotted line), and where they could head if recent policy pronouncements and shifts in trends (*e.g.* the US shift away from MTBE and towards ethanol) were sustained.

Figure 7.1

Fuel Ethanol Production, Projections to 2020

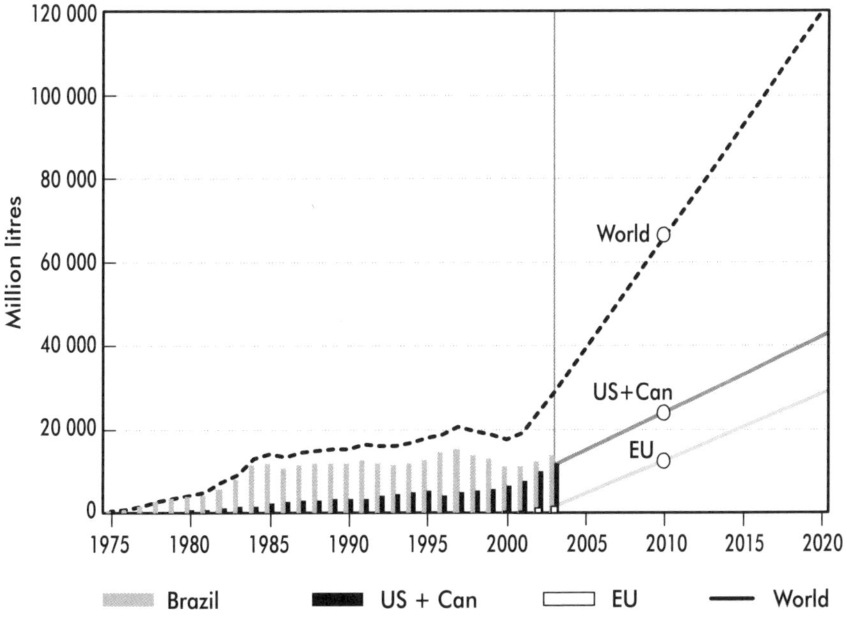

If historical trends were to continue (not shown), annual growth rates in the future would be about 7% for Europe, 2.5% for North America and Brazil, and 2.3% for the whole world. This would lead to a global increase from about 30 billion litres in 2003 to over 40 billion by 2020. However, given recent policy initiatives and changes in trends, a very different picture could emerge: a quadrupling of world production to over 120 billion litres by 2020 (see Figure 7.1). On a gasoline energy-equivalent basis, this represents about 80 billion litres, or nearly 3 exajoules. This would likely account for about 6% of world motor gasoline use in 2020, or about 3% of total road transport energy use[8].

Various targets and factors have been taken into account to develop this "alternative" projection. These include the EU target for 2010, the proposed US target for a doubling of ethanol use in the 2010 time frame, and various other announced targets and new initiatives discussed in the previous section. The world "target" for 2010 (noted by a dot on the upper-most line) is simply the sum of various targets and initiatives from around the world. This higher trajectory is then carried through to 2020 to illustrate where it would lead.

A similar projection was undertaken for biodiesel. Here as well, the contrast between existing trends and a target-linked trend is stark, though it is mainly attributable to one programme – the EU voluntary targets for biofuels. It assumes that under the EU directive, countries will choose to meet their 5.75% transport fuel displacement commitments proportionately with biodiesel (for diesel) as with ethanol (for gasoline). If this occurs, it will result in more than a tenfold increase in biodiesel production in the EU. However, as discussed in Chapter 6, biodiesel from FAME requires more land per delivered energy than ethanol, and some countries may choose to displace gasoline (with ethanol) more than proportionately, and diesel (with biodiesel) less than proportionately.

While achieving these new, higher trajectories would require large investments and increases in biofuels production, it is still less than what appears possible on a global basis (see Chapter 6). The study by Johnson (2002) suggests that over 7% of road transport fuel (10% of gasoline, 3% of diesel) could be displaced with cane ethanol alone in the 2020 time frame. Most of the

---

8. *The WEO 2002 projects world transport energy use of 110 exajoules in 2020, or about 90 for road transport (at the same share as in 2000).*

Figure 7.2

Biodiesel Production Projections to 2020

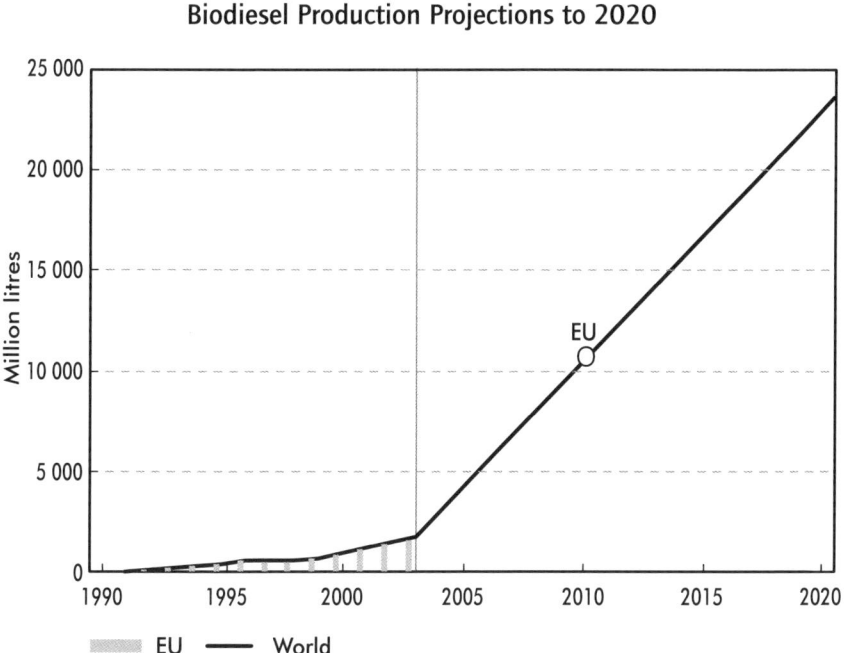

production targets and initiatives described above are based on domestic production, though some trade is anticipated. As demand for biofuels rises, at some point, for some countries, domestic production is likely to reach certain limits, or trigger unacceptable costs. Thus, in order to continue on a rapid growth path after 2020, new approaches will likely be needed, which could include targeting increases in production in the most suitable, lowest-cost regions, and expanding global trade. Chapter 8 looks at, among other things, the current situation and potential barriers to trade.

8 BENEFITS AND COSTS OF BIOFUELS AND IMPLICATIONS FOR POLICY-MAKING

As described throughout this book, displacing gasoline and diesel fuels with liquid biofuels for transport brings considerable benefits, such as improving energy security, protecting the environment, and enhancing agricultural productivity. Considering these benefits, various governments within and outside the OECD have advanced policies to support the development and deployment of biofuels – from basic research to mandates on fuel and vehicle use. Many governments are in the process of implementing ambitious measures; some are initiating dialogue on introducing biofuels; and still others are watching and evaluating progress and lessons learned.

Regardless of the level and type of current involvement, policy-makers are grappling with several key issues:

- How do the benefits of biofuels compare with the costs?
- Which policy measures should be pursued if net benefits are to be maximised?
- Where should future research and development work be focused?

This chapter draws on the material presented throughout this book to help answer these questions, beginning with a review of the benefits and costs associated with an expanded use of biofuels, and their potential for displacing petroleum fuels. The second section lays out a series of traditional and non-traditional policy measures for promoting biofuels. And, finally, the chapter and this book end with a view towards the future, by suggesting areas where additional analytical and technical research on biofuels is needed.

The Benefits and Costs of Biofuels

Increasing the use of biofuels would yield net benefits both locally and globally. However, the benefit-cost evaluation is dominated by several difficult-to-quantify benefits, while costs are dominated by the fairly well-quantified – and often fairly high – production costs. Estimating the value of the benefits is one of the most difficult and uncertain aspects of biofuels analysis But without such an analysis, there is a tendency to focus on costs. For example, the cost per tonne of CO_2 emissions reductions using conventional biofuels in OECD countries, given current feedstock and conversion technology, appears to be high. However, this measured cost might be much lower when considered in the context of other not-yet-measured benefits, such as improvements in energy security, reductions in pollutant emissions, fuel octane enhancement and improvements in the balance of trade. The various benefits and costs that need weighing, and that have been considered here, are shown in Table 8.1.

Table 8.1

Potential Benefits and Costs of Biofuels

Potential benefits	Potential costs
• Energy security	• Higher fuel costs
• Balance of trade	• Increases in some air emissions
• Lower GHG emissions	• Higher crop (and crop product) prices
• Reduced air pollution emissions	• Other environmental impacts, such as land use change and loss of habitat
• Vehicle performance	
• Agricultural sector income, jobs and community development	
• Waste reduction	

To date, much of the research on biofuels has focused on monetising only a few of these impacts. Apart from fuel production costs, values for the benefits or costs to society of most of the other items in Table 8.1 have not been systematically quantified. In some cases, such as for monetising the value of GHG emissions reductions, considerable work has been done, but a wide range of estimates exists.

It is therefore very important for governments to undertake the job of better estimating these benefits and costs. The monetary values of reducing oil use and of lowering greenhouse gas emissions are particularly important, but they are also uncertain and controversial. In part, this explains why it is difficult in many countries to implement policies that would help alleviate these problems (such as taxes to "internalise" the social costs of petroleum fuel use and dependence). Another area of uncertainty and complexity is the macroeconomic impacts of biofuels use – for example, factoring in the effects of diverting crops towards biofuels production on other markets (*e.g.* food) and balancing this against benefits (*e.g.* increased income for farmers and rural areas). Similarly, employment benefits in areas producing biofuels need to be compared to possible job losses in other regions or sectors.

Unless societies make an effort to carefully estimate the value of the key costs and benefits associated with biofuels use, decisions about whether and how much to produce will likely be dictated more by sectoral interests and political expediency than by an effort to maximise overall social welfare. In the literature reviewed in this book, large information gaps have been found. An important follow-up area is to better quantify each of the benefits and costs outlined below.

Improved Energy Security

Governments have long sought to reduce petroleum import dependence, primarily to improve energy security and the balance of trade. However, there are few agreed methods for evaluating energy security or quantifying the cost of insecurity. Without such methods, it is difficult to make clear cost-benefit trade-offs between financial, technical and policy measures intended to improve energy security or to evaluate energy security measures in a larger policy context. The "Herfindahl" measure of market concentration is one possible indicator. This measure relates market size to risk dependency. The greater the number of suppliers or fuel supplies, the lower the risk dependency (Neff, 1997). As such, if a country were dependent on petroleum from one country (or region) for 95% of transportation needs, it would have an associated dependency index of 0.90. If biofuels (produced domestically, or imported from a different region) were to replace 10% of the petroleum in this market, the dependency index would fall to 0.74. Using this measure, the benefit of diversification is the same even if biofuels are imported, as long as they come from countries or regions other than those supplying oil. This would be the case for biofuels like ethanol produced in countries such as Brazil and India.

Improved Balance of Trade

Oil accounts for a significant percentage of total import costs for many countries. For example, the US imported $106 billion of crude oil and petroleum products in 2000 (some 48% of consumption), accounting for almost one-third of the total US trade deficit in goods and services (US Census, 2001). Increasing the share of domestically produced biofuels in the US transportation market to 10% would reduce oil consumption by about 8% (given that some oil is used to make ethanol). If all of the oil reduction came from imports, oil imports would drop by about 15%, saving over $15 billion in import costs. Generally speaking, a lower trade deficit will benefit the macro-economy, by spurring domestic economic activity. Unlike energy security, the trade balance would not benefit from substituting imported biofuels for imported petroleum fuels – unless biofuels were cheaper.

Reduction in Greenhouse Gas Emissions

On a global scale, vehicle emissions contribute nearly 20% of energy-related greenhouse gas emissions (IEA, 2002). Including upstream emissions associated with fuel production, the percentage is closer to 25%. As discussed in Chapter 3, both ethanol and biodiesel can provide significant "well-to-wheels" reductions in greenhouse gas emissions compared to gasoline and diesel fuel. Studies reviewed for this book indicate up to a 40% net reduction from grain ethanol versus gasoline, up to a 100% reduction from cellulosic and sugar cane-derived ethanol, and up to a 70% reduction from biodiesel relative to diesel fuel.

The value of these reductions depends on what impact these gases ultimately have on the atmosphere, and how much damage this impact causes. Given the high uncertainty, estimates of this type are quite speculative and most researchers avoid making them, instead simply comparing the costs of various measures for reducing GHG emissions and recommending adoption of the lowest-cost measures. When looked at in isolation, the cost per tonne of GHG reductions from biofuels is quite high – at least for biofuels produced in OECD countries – ranging up to $500 per tonne of CO_2-equivalent GHG reduction (as discussed in Chapter 4). However, if other benefits and costs of biofuels are taken into account, and the remaining net cost is compared to GHG reduction, the cost per tonne is likely to be much lower, and possibly, under some circumstances, negative (*i.e.* the use of biofuels provides net benefits apart from GHG reduction).

Reduction in Air Pollution

As discussed in Chapter 5, biofuels can provide certain air quality benefits when blended with petroleum fuels, particularly in urban areas. They (particularly ethanol) may also increase emissions of certain pollutants. Measurement and evaluation are also affected by the types of emissions controls on vehicles. As controls are tightened, the direct effects of biofuels diminish, though the cost of compliance to meet a certain standard may decrease. Benefits from ethanol blending include lower emissions of carbon monoxide (CO), sulphur dioxide (SO_2) and particulate matter (PM). Benefits from biodiesel include all these plus lower hydrocarbon emissions. Biofuels are also generally less toxic than conventional petroleum fuels. Ethanol use may lead to increased aldehyde and evaporative hydrocarbon emissions and both ethanol and biodiesel may cause small increases in NO_x, particularly on a well-to-wheels basis. Estimating the net benefits of these changes in emissions is complex, as they differ by country, city, time of year, etc. No studies have been found that monetise the costs and benefits of pollutants from biofuels for a particular region or country. Net benefits could be substantial, particularly from PM reduction in cities with high average tailpipe emissions, such as in many developing countries.

Improved Vehicle Performance

As discussed in Chapter 5, biofuels can provide significant vehicle performance benefits. Biodiesel can significantly improve the performance of conventional diesel fuel even when blended in small amounts (*e.g.* B5). Ethanol has a high octane number and can be used to increase the octane of gasoline. It has not traditionally been the first choice for octane enhancement due to its relatively high cost, but as other options become increasingly out of favour (leaded fuel is banned in most countries and MTBE is being discouraged or banned in an increasing number of countries), demand for ethanol for this purpose, and as an oxygenate, is on the rise in places such as California. In Europe, ethanol is typically converted to ethyl-tertiary-butyl-ether (ETBE) before being blended with gasoline. ETBE provides high octane and oxygenation with lower volatility than ethanol, but is 60% non-renewable isobutylene.

From the point of view of vehicle performance, the marginal cost of ethanol probably should be viewed as its opportunity cost – taking into account the cost and characteristics of the additive(s) it replaces. It may be the case that after taking into account ethanol's octane, oxygenation and emissions benefits, its net cost (per unit vehicle-related benefit) is not much higher than additives like MTBE; and it is certainly of lower social cost than lead (Pb), taking into account the high cost of lead's impact on health. For low-level blends of ethanol and biodiesel, comparing their costs to other fuel additives that provide similar benefits may make more sense than comparing to the cost of base fuels (gasoline and diesel), and also helps quantify their vehicle-performance benefits. More work is needed in this area.

Agro-economy

Some governments have been attracted to the potential role biofuels can play in stimulating domestic agricultural production and expanding the markets for domestic agricultural products. Production of biofuels from crops such as corn and wheat (for ethanol) and soy and rape (for biodiesel) provides new product market opportunities for farmers, with the potential to increase farming revenues or expand the productive capacity of existing cropland.

As shown in Chapter 6, fuel ethanol production sufficient to displace 5% of gasoline could require approximately 30% of the US corn crop and 10% of EU production of wheat and sugar beet. The discussion in Chapter 4 of the impacts of biofuels production on crop prices suggests that crop prices typically rise when new markets for them are created (since demand increases while supply does not, at least initially, change). This creates a wealth transfer from consumers of these crops to producers (farmers). This can provide important benefits to rural economies, a priority for many governments.

Another potentially important dynamic is the impact of increased crop production on existing subsidy payments. In both the US and the EU, certain programmes compensate farmers for set-aside land. If this land could be used to grow crops for biofuels in an eco-friendly manner that preserves this often sensitive land, then existing subsidies could be retargeted towards more productive activities. As discussed in Chapter 7, the recent EU common agricultural policy (CAP) reforms are moving in this direction, and some use of set-aside land for biofuels crop production was already allowed. However, there may be opportunities in the EU and elsewhere for much stronger efforts to promote production of environment-friendly crops such as switchgrass, along with production of more environment-friendly biofuels (*e.g.* low-greenhouse-gas ethanol from cellulosic feedstocks). This is an area where more research – and policy reform – is needed.

Impacts on Markets and Prices

While the impact of increased biofuels production on farm income is expected to be mainly positive, the net effect on all groups is much less clear. For example, diversion of crops to produce biofuels is likely to cause a rise in other crop (and crop product) prices due to lower availability. However, as mentioned above and discussed in Chapter 4, given the current approach to subsidising farm production in many countries, it is much more difficult to estimate the impact, at the margin, of increasing biofuels production. In some cases it could divert existing subsidies to a new activity, which might be more productive than the current impact of the subsidy (such as if the subsidy encourages farmers to set aside land). There is also significant overproduction of some crops in many IEA countries, and the development of new markets may be able to absorb existing oversupply before drawing crops away from other purposes. This area of analysis deserves much greater attention than it has received to date.

Waste Reduction

As discussed in Chapter 5, biofuels can reduce certain types of organic wastes through recycling – including crop waste, forestry wastes, municipal wastes, and waste oils and grease which can be converted to biofuels. Much of the world's waste products are cellulosic in nature (*e.g.* wood, paper and cardboard). Municipalities dispose of tonnes of paper and yard waste. Some segments of the agricultural and forest products industries produce huge amounts of lignocellulosic waste. Other co-mingled wastes amenable to biofuels production include septic tank wastes, wastewater treatment plant sludge (so called biosolids), feed lot wastes and manure. A large proportion of uncollected (primarily household) waste oil is likely being dumped into sewage systems or landfill sites, even though it is illegal in many jurisdictions where waste oil is a listed waste substance. A number of studies conducted in the EU over the last several years point to a possible supply of readily collectible waste cooking oil and grease exceeding one million tonnes, which could be used to produce around one billion litres of biodiesel, or about two-thirds of biodiesel production in the EU in 2002 (Rice *et al.*, 1997).

Reduction and redirection of these waste streams towards productive uses clearly provides a social benefit. To some extent, this benefit can be measured by the avoided cost of otherwise disposing of the waste. For example, the government in the UK currently levies a tax of £14 (about $24) per tonne of landfill waste, based on an estimation of the direct and indirect social costs. The value per litre of ethanol or biodiesel depends on how much is produced per tonne of waste, which in turn depends on the properties of this waste. For biodiesel, a little more than 1 000 litres can be produced from a tonne of waste oil. This translates into some $0.02 of savings per litre of biodiesel produced from waste oil. Because of the avoided disposal costs, the feedstock cost for biofuels production would essentially be negative (-$0.02). In Chapter 4, biodiesel costs are shown to be much lower if produced from waste oil or grease than from oil-seed crops.

Higher Fuel Prices

As discussed in Chapter 4, though biofuel production costs have dropped somewhat over the past decade, conventional (grain) ethanol and biodiesel produced with current technology in OECD countries are still two to three times more expensive than gasoline and diesel. In some developing countries, it appears that ethanol from sugar cane is competitive – or close to it – with imported petroleum fuel. Estimates of the production cost of biofuels from particular conversion facilities or for a particular country are fairly easily obtainable, so this is one item in Table 8.1 that is well quantified.

Still, there may be hidden (or not so hidden) taxes or subsidies in estimated production costs and market prices. For example, given the complex agricultural policies in places like the US and the EU, crop prices are likely to be quite different from their true marginal production costs. Of course, petroleum prices also tend to depart dramatically from their marginal production costs.

Fuel-Vehicle Compatibility

The cost of making vehicles compatible with biofuels is particularly difficult to measure, because it is difficult to define. The main criterion is whether the use of any particular blend level requires modification of vehicles, or causes problems for vehicles if no modifications are made. In most countries, blends are capped at levels that are believed to avoid causing any vehicle problems. In this case, the primary consideration is the costs associated with making the vehicles compatible with the fuel. As mentioned in Chapter 5, for 10% blends only very minor modifications are required to vehicles and most manufactures have already made these modifications to vehicles sold in parts of the world where blending occurs. These costs may be on the order of just a few dollars per vehicle. In countries like Brazil, where higher blend levels are used, vehicle costs are higher. Experience in the US with flex-fuel vehicles indicates that vehicles can be made compatible with up to 85% ethanol for a few hundred dollars per vehicle. This cost is likely to come down over time, with technological improvements and with mass production. Biodiesel blending with diesel appears to require few or no modifications to diesel engines.

Policies to Promote Increased Use of Biofuels

As shown in Chapter 7, biofuels production in IEA countries is growing rapidly. However, given the currently high production cost of biofuels compared to petroleum fuels in these countries, it is clear that much of this increase is driven by new policies. It is unlikely that biofuels use will grow rapidly in the future without continuous policy pressure. Since many countries are still considering how best to promote biofuels, this section discusses a variety of traditional and non-traditional policy approaches.

Fuel Tax Incentives

Typically, the most daunting aspect to the use of biofuels (*e.g.* for refineries, as an octane enhancer) is the purchase price. Fuel tax incentives can therefore be a very effective tool for encouraging the use of biofuels, making them more price-competitive with petroleum fuels (and with competing octane enhancers, oxygenates, etc.). These incentives can be especially effective during the early years of fuel market development, if costs are expected to come down as the scale and experience of biofuel production increases (*i.e.* in Brazil). Since fuel excise taxes comprise a significant percentage of the price consumers pay for motor fuels, particularly in Europe and Japan, exempting alternative fuels from a portion of this tax burden is an available and powerful tool for "levelling the playing field". This incentive also sends a clear signal to consumers regarding the relative social costs of different fuels. If the externalities associated with the use of biofuels are lower than those of petroleum fuels, a lower tax on biofuels is economic.

One common concern about setting a lower tax for biofuels (and other alternative fuels), however, is that it will reduce government revenue. This can be avoided by adjusting the taxes on all fuels so that total revenues are maintained. Tax rates would have to be modified periodically to adjust to changes in demand for each fuel. But it is often difficult for legislatures to frequently change tax rates.

Carbon-based Fuel Taxes

Carbon taxes are fuel taxes based on the carbon content of the fuel. Carbon taxes make sense economically and environmentally because they tax the externality (carbon) directly. They can be an effective stimulant for alternative fuels (and alternative-fuel vehicles) in cases where lower emissions result in a significantly lower levied tax rate. However, while carbon-based fuel taxation is relatively straightforward, for biofuels to appear attractive it would be necessary to develop a scheme that takes into account well-to-wheels emissions, not just tailpipe emissions. This is a complex undertaking, because the scheme would vary considerably depending on how biofuels (and other fuels) are produced.

Many countries have variable fuel or vehicle taxes based on carbon content or CO_2 emissions per kilometre driven. Sweden, Finland, Norway, the Netherlands and Slovenia tax fuels on the basis of their carbon content. But no country is known to take into account upstream emissions. In the case of biofuels, strong differentiation of fuel tax (or subsidy) based on well-to-wheels GHG emissions will serve to promote new, more environment-friendly biofuels such as cellulosic ethanol and biomass-to-liquids (BTL) via gasification with Fischer-Tropsch processes, hydrothermal upgrading (HTU) processes, etc. As discussed in Chapter 6, such advanced biofuels also will allow a broader base of feedstocks to be used, with better conversion efficiencies, thus increasing potential supply. Governments could therefore substantially increase the overall social benefits of biofuels use through differential taxation of biofuels based on process and GHG characteristics.

Vehicle Taxes and Subsidies

In addition to fuel-related incentives, fuel consumption can be affected by policies which encourage the purchase of vehicles running on certain types of fuel, or running on fuels that emit less CO_2. Denmark, the Netherlands and the UK have recently introduced new vehicle tax rates based at least in part on CO_2 emissions (though the Netherlands suspended their scheme after one year). For example, in the UK, the base vehicle registration fee is set at 15% for vehicles emitting 165 grams of CO_2 per kilometre driven. For each 5 grams additional CO_2 (depending on the rated fuel economy of the car), an additional one percentage point is added to the tax. For diesel, 3 percentages points are added. However, this approach provides little incentive to use biofuels since they have little effect on vehicle emissions of CO_2. The scheme would have to take into account upstream CO_2 for biofuels to receive a tax break.

CO_2 Trading

Under an emissions trading system, the quantity of emissions allowed by various emitters is "capped" and the right to emit becomes a tradable commodity, typically with permits to emit a given amount. To be in compliance, those participating in the system must hold a number of permits greater or equal to their actual emissions level. Once permits are allocated (by auction, sale or free allocation), they are then tradable.

A well-functioning emissions trading system allows emissions reductions to take place wherever abatement costs are lowest, potentially even across international borders. Since climate change is global in nature and the effects (*e.g.* coastal flooding, increasing incidence of violent storms, crop loss, etc.) have no correlation with the origin of carbon emissions, the rationale for this policy approach is clear. If emissions reductions are cheaper to make in one country than another, emissions should be reduced first in the country where costs are lower.

Emissions trading systems could include biofuels and create an incentive to invest in biofuels production and blending with petroleum fuels (*e.g.* by oil companies) in order to lower the emissions per litre associated with transport fuels, and reduce the number of permits required to produce and sell such fuel. However, as for tax systems, in order for biofuels to be interesting in such a system, the full well-to-wheels GHG must be taken into account.

Clean Development Mechanism (CDM) and Joint Implementation (JI)

Under the Kyoto Protocol, countries can engage in projects through which an entity in one country partially meets its domestic commitment to reduce GHG levels by financing and supporting the development of a project in another country. JI projects are between two industrialised countries. CDM projects are between an industrialised and a developing country. In both cases, one country provides the other with project financing and technology, while receiving CO_2 reduction credits that can be used in meeting its emissions reduction commitments. A major requirement for CDM projects is that they also have to further the sustainable development goals of the host country. In addition, CDM projects must involve activities that would not otherwise have occurred, and should result in real and measurable emissions reductions. The two most common types of projects tend to be land use and energy – which demonstrate potentials for biofuels (*i.e.* crops planted in exchange for energy-related vehicle emissions reductions). For this reason, there is an increasing awareness of the opportunities for producing biofuels from community-scale plantations in developing countries.

An example of a CDM project is that between Germany and Brazil, where Germany will purchase carbon credits from Brazil as part of its Kyoto Protocol commitments. In turn, the funds will help Brazil subsidise taxi use of biofuels and the development of dedicated ethanol vehicles. It is not yet clear how the well-to-wheels GHG savings will be measured.

Argentina, one of the world's biggest producers and exporters of oil-seeds, has also expressed interest in using the CDM for maximising its enormous potential for biodiesel production. The government is hoping that the CDM can offer a triggering incentive to encourage producers and investors to develop project activities and it has established institutional support for biodiesel-oriented CDM projects by creating the Argentine Office of the Clean Development Mechanism (EF, 2002). The same office is also co-ordinating the Biofuels National Programme, which aims to promote the production and use of biofuels.

Despite this promising tool for stimulating biofuels production and use, there are several reasons why only recently the first GHG emissions reduction projects involving biofuels and transportation have emerged. One is the limited experience and methodologies for estimating, monitoring and certifying potential well-to-wheels emissions reductions from transport projects. This is changing quickly, with the proliferation of well-to-wheels GHG assessments, though there still are very few studies for non-IEA countries.

A related reason is the lack of a commonly agreed CDM/JI methodology and data for estimation of emissions baselines for this type of project. In general, a fuel-switching project should not pose particularly difficult baseline-measurement issues, but, as mentioned, tracking the emissions from all upstream fuel production-related activities is difficult, and any required change to vehicles complicates matters somewhat. However, several recent CDM projects have been approved that focus on biomass-to-electricity generation, and an extension of this methodology to biofuels production (or co-production with electricity) should be possible.

As for all sectors, projects will not have value until a market develops where emissions reduction credits have a tradable value. The Kyoto Protocol should provide this but as of January 2004 it is not yet ratified. Even when ratified, it may take many years before some countries find it necessary to turn to relatively expensive transport projects for CO_2 credits. If cost per tonne of GHG reduction can be brought down well below $50, as appears to be occurring in Brazil, this will certainly make biofuels projects more attractive.

Fuels Standards

Governments can also implement fuel standards as a mechanism for altering the transport sector fuel mix. Many governments already use fuel quality standards to help protect public health and the environment from harmful gaseous and particulate emissions from motor vehicles and engines, and to help ensure compatibility between fuels and vehicles. Such standards have included a gradual phasing-out of lead to reduce the health risks from lead (Pb) emissions from gasoline; measures to reduce fuel volatility so as to mitigate ozone, particularly in summer months; and standards which gradually reduce the level of sulphur content in fuels. By implementing a standard for minimum fuel content of non-petroleum (or renewable) fuel, governments could similarly use regulation to drive the market. This approach has the advantage of clearly defining the market share reserved for specific types of fuels, such as biofuels. It creates a stable environment to promote fuel production and market development. A disadvantage of this approach is that costs are uncapped, *i.e.* fuel providers must comply regardless of costs.

Incentives for Investment into Biofuels Production Facilities

Apart from fuel-related incentives, an important barrier to the development of a market for biofuels is the required investment in commercial scale production facilities. Fuel providers have little incentive to make large investments in these facilities in the current uncertain market. Even if governments put into place fuel incentives that generate demand for the fuel, investors will be wary that such policies can change at any time. In order to encourage the necessary investment, governments may consider certain investment incentives such as investment tax credits or loan guarantees.

Trade Policy to Remove Barriers to International Biofuels Trade

Given the wide range of biofuels production costs worldwide (as shown in Chapter 4) and the wide range in production potential for biofuels in different countries (as shown in Chapter 7), there appears to be substantial potential benefits from international trade in biofuels. However, at present, there is no comprehensive, nor is there even a substantial specific, trade regime applicable to biofuels. Biofuels are treated either as "other fuels", or as alcohol (for ethanol) and are subject to general international trade rules under the WTO (*e.g.* Most Favoured Nation Principle; National Treatment; general elimination of quantitative restrictions; prohibition of certain kinds of subsidies, etc.).

Failing specific rules, biofuels are generally subject to customs duties and taxes without any particular limits. These tariffs vary substantially from one country to the other. The ethanol market in several developed countries is strongly protected by high tariffs, and OECD countries apply tariffs of up to $0.23 per litre for denatured ethanol (Figure 8.1). Some countries also apply additional duties to their tariffs, *e.g.* the US applies *ad valorem* tariffs of 2.5% for imports from most-favoured-nation (MFN) countries and 20% for imports from other countries. Japan applies *ad valorem* tariffs of 27% (MFN treatment).

Given that ethanol produced in countries like Brazil appears to be on the order of $0.10 to $0.20 per litre cheaper to produce than in IEA countries (as discussed in Chapter 4), and that ocean transport costs are probably less than a penny per litre, duties on the order of $0.10 per litre or higher represent a significant barrier to trade.

However, ethanol is included in a list of environmental products for which accelerated dismantling of trade barriers is sought, so there are some prospects for the eventual elimination of these tariffs (see box).

Vehicle Requirements for Compatibility

A non-traditional policy tool available to governments could be the introduction of vehicle technology standards that require compatibility with

BIOFUELS FOR TRANSPORT

Erratum for Figure 8.1, p 185

Since the publication of this book, a small error has been discovered on figure 8.1 - Ethanol Import Duties Around the World.

The following provides the updated, corrected version of this figure. We apologise for any inconvenience caused.

Figure 8.1

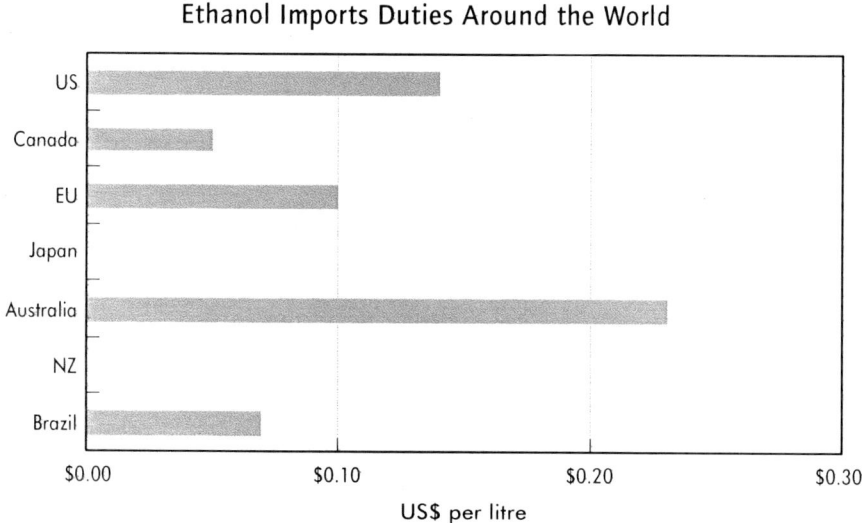

Ethanol Imports Duties Around the World

Note: Ethanol import duties in Japan and New Zealand are zero.
Source: Various national tax reports and websites.

Figure 8.1

Ethanol Import Duties Around the World

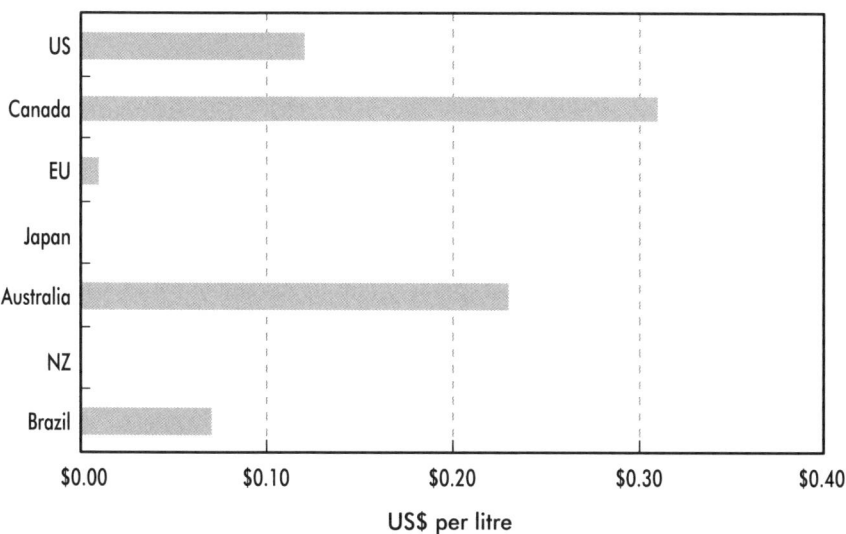

US$ per litre

Note: Ethanol import duties in Japan and New Zealand are zero.
Source: Various national tax reports and websites.

Recent WTO Initiatives Affecting Biofuels

At the Doha Ministerial meeting of the WTO in Cancun, September 2003, the declaration called for negotiations on "the reduction or, as appropriate, elimination of tariff and non-tariff barriers to environmental goods and services". However, the term "environmental goods" was not defined in the declaration. A substantial amount of work to identify the scope of environmental goods has already been undertaken by the OECD and APEC (Asia-Pacific Economic Co-operation), culminating in two product lists of candidate goods (OECD, 2003). Both lists contain ethanol, classified under the HS 220710 (OECD, 2003). Negotiations will continue and biofuels may be included in future lists of environmental goods and services for which tariff reductions are negotiated.

specific mixtures of biofuels. Brazil has essentially done this through a fuel standard, requiring all gasoline to be blended with 22% to 26% ethanol. This has forced manufacturers to ensure that their vehicles are compatible with these blends. In the US, and now in Brazil, several manufacturers have introduced flexible fuel capability in a number of vehicle models. Such vehicles can run on low or high-level ethanol blends, and the conversion cost (estimated at no more than a few hundred dollars per vehicle) is included in the vehicle price. If all new vehicles were required to be at least E0-E85 compatible, then ethanol could be used in any vehicle in any part of the world. Further, if all vehicles produced were of this type, the costs for producing such vehicles would probably drop considerably due to scale economies – perhaps to less than US$ 100 per vehicle above non-flex-fuel versions.

Areas for Further Research

Some important areas for needed additional research into biofuels are outlined below.

Increased R&D for Cellulose-to-Ethanol and Other Advanced Processes

Since ethanol can be produced from any biological feedstock that contains appreciable amounts of sugar or materials that can be converted into sugar such as starch or cellulose, a key research goal is to develop cellulosic conversion technologies. In a few countries, research efforts are already well under way to develop methods to convert cellulosic materials to ethanol (by first breaking the cellulose down into sugars). These efforts are promising for several reasons: *i)* a much wider array of potential feedstock (including waste cellulosic materials and dedicated cellulosic crops such as grasses and trees), opening the door to much greater ethanol production levels; *ii)* a much greater displacement of fossil energy, due to nearly completely biomass-powered systems; and *iii)* much lower well-to-wheels greenhouse gas emissions than is currently the case with grain-to-ethanol processes. These conversion processes are also potentially low-cost, and as mentioned in Chapter 4, some studies estimate that cellulosic ethanol could become cheaper than conventional ethanol in the 2010-2020 time frame. Though other approaches for converting biomass into biofuels for transport are also promising, most are not expected to reach cost-competitive levels as soon as cellulosic ethanol.

While the US and Canada are putting considerable resources into cellulosic ethanol research, many other countries are not investigating this option. Progress has been fairly slow in recent years. No large-scale test facilities have yet been built (though several are now planned). In order to optimise the technology to achieve economies of scale and to generate the kind of learning-by-doing that drives down costs, many more countries need to become involved in financing research and development. In addition, efforts to construct commercial-scale facilities need to be intensified.

On the other hand, many countries (particularly in Europe) are putting considerable resources into other possible approaches to converting cellulose and other forms of biomass into biofuels (Chapter 2). These approaches typically involve biomass gasification and conversion to various fuels, including synthetic diesel and gasoline. Many are promising and deserve greater attention. The IEA Bioenergy Implementing Agreement helps co-ordinate much of the research in this area and would benefit from greater involvement and support from IEA member and non-member countries alike (non-IEA member countries may join IEA implementing agreements).

Land Use Impacts, Costs, and Global Production Potential

Though Chapter 6 provides some general estimates of how much land could be required to produce different amounts (and different types) of biofuels, it is clear that much more research is needed to better understand this very important area. For example, most available estimates have not attempted to estimate how much crops and other types of feedstock might become available for biofuels production at different prices. This is particularly true for global analyses. There no doubt exists a feedstock "supply curve", and possibly a fairly steep one, that could affect biofuels production costs regionally and, if strong international trade emerges, globally. While low-cost production of biofuels appears possible around the developing world, there is a poor understanding of how costs would change if production were expanded dramatically – possibly into less productive or more expensive types of land.

Similarly, as discussed in Chapter 4, competition between feedstock production for biofuels and for other purposes can affect costs and prices significantly. As feedstock is drawn away from other purposes for biofuels production, costs of other products may rise. This can be a very good thing for producers (like farmers), but not necessarily for consumers or for society as a whole. Market equilibrium impacts of new policies need to be better understood, and considered more often, than they currently are.

Finally, in terms of global production potential, most studies appear to have allocated land in descending order of importance – and value – for example by calculating how much land is required for cities and other areas of human habitation, food production, and conservation areas, before estimating how much of the residual might be suitable for different types of biofuels production. While this approach is sensible, it may ignore important opportunities for high-efficiency co-production of different materials – *e.g.* ethanol, electricity and feed grains from cellulosic crops. At least some observers (*e.g.* Lynd, 2004) believe that existing studies significantly underestimate global biofuels production potential by ignoring such opportunities. Such opportunities may also help to alleviate the problem of feedstock competition for different uses, and keep costs down.

Interactions between Agricultural Policy and Biofuels Production

Although a discussion of agricultural policy is mostly outside the scope of this volume, several ways in which agricultural policy and biofuels production can interact have been discussed at various points. Most IEA countries, and the EU, have complex agricultural policies that make it difficult to understand what impact increased biofuels production would have on things like crop prices, agricultural subsidies, and net social welfare. As noted in Chapter 4, subsidies to farmers to produce biofuels may, in some cases, help to offset other subsides – for example, in the US there are programmes to assist farmers if crop prices fall below certain levels. With additional demand for crops for biofuels production, this might happen less frequently.

With major reforms to agricultural policy under consideration in the EU as well as in various IEA countries, along with initiatives for substantially increased production of biofuels, a better understanding is needed of how policy in these two areas interacts, and how policy could be optimally designed in this regard. It may be possible, for example, to convert some existing subsidies that encourage farmers not to plant (in order to help maintain crop price levels) into subsidies that encourage the production of crops for biofuels production.

While such policy shifts have been discussed, and even implemented in a few cases, they deserve more attention. Agricultural policies could also be used to encourage the most environment-friendly approaches to biofuels production, such as the use of switchgrass in environmentally sensitive areas.

Research into Net Costs and Benefits of Biofuels

Though it may be possible to shift existing subsidies to encourage production of biofuels, ultimately the question becomes whether biofuels should be subsidised at all, or not. There is a case for a long-term subsidy, if biofuels provide net societal benefits that are not captured in the market system. Currently, looking only at production cost, biofuels (at least those produced in IEA countries) seem expensive as options for reducing greenhouse gases. But no available study has taken the much broader view of attempting to assess, quantitatively, the costs and benefits in the many areas discussed at the start of this chapter, and throughout this book. There is a strong need for objective, detailed research in this area. Though point estimates of things like air quality impacts will always be difficult to make (since there can be widely varying impacts depending on the specific situation – vehicle type, emissions control, geography, ambient conditions, etc.), approximate estimates may be all that is needed in order to better gauge whether, and under what conditions, using biofuels provides net benefits to society and how these benefits can be maximised.

A related aspect deserving research is how net costs and benefits are likely to change as production of biofuels increases. As mentioned above, a production cost curve is needed for biofuels production worldwide, since costs vary considerably by region, and perhaps by scale. But other types of costs and benefits are also likely to vary by overall production scale. For example, as mentioned in Chapter 4, if biodiesel production is increased very much in any given region, the market for the co-product glycerine is likely to be saturated quickly and additional glycerine would likely have little value, thus effectively increasing the cost of biodiesel. As discussed in Chapter 5, certain types of vehicle-related costs go up as compatibility is sought for higher blend levels of ethanol. Most regions can go to 10% gasoline displacement with ethanol (on a volume basis), without any vehicle-related costs, but going above this level entails some additional (vehicle conversion) cost. A study that develops a cost/benefit curve, covering current levels of production and use, and on up to much higher levels, would be welcome.

ABBREVIATIONS AND GLOSSARY

BTL   biomass-to-liquids

Bxx   (where xx is a number, *e.g.* B5, B10, etc.) biodiesel blend with petroleum diesel, with biodiesel volume percentage indicated by the number

CAP   Common Agricultural Policy (EU)

CBP   combined bioprocessing (technology for cellulosic ethanol production)

CCs   EU candidate countries

CDM   Clean Development Mechanism (under Kyoto Protocol)

CH_4 methane

CNG   compressed natural gas

CO   carbon monoxide

CO_2 carbon dioxide

DDGS   distillers dry grains soluble

DME   dimethyl ether

DOE   Department of Energy (US)

EC   European Commission

E-diesel   ethanol-diesel blends

EPA   Environmental Protection Agency (US)

ETBE   ethyl tertiary butyl ether

EU   European Union

Exx   (where xx is a number, *e.g.* E10, E20, etc.) ethanol blend with gasoline, with ethanol volume percentage indicated by the number

FAME   fatty acid methyl ester (biodiesel)

FFV   flexible-fuel vehicle

F-T   Fischer-Tropsch (process for making synthetic fuels)

GE	genetic engineering (or genetically engineered)
GHG	greenhouse gas
GMO	genetically modified organisms
GWP	global warming potential
H_2	hydrogen
HC	hydrocarbons
HTU	hydrothermal upgrading
JI	Joint Implementation (under Kyoto Protocol)
Kcs	Czech kroner (currency)
LPG	liquefied petroleum gas
MFN	most-favoured nation (under WTO)
MJ	megajoule
MTBE	methyl tertiary butyl ether
N_2O	nitrous oxide
NMHC	non-methane hydrocarbons
NMOC	non-methane organic compound (similar to NMHC)
NO_x	oxides of nitrogen
NREL	National Renewable Energy Laboratory (US)
O_3	ozone
OECD	Organisation of Economic Co-operation and Development
PAN	peroxyacetyl nitrate
PM	particulate matter
ppm	parts per million
R$	Brazilian real (currency)
R&D	research and development
RFG	reformulated gasoline
Rs	Indian rupees (currency)
RME	rapeseed methyl ester (a type of FAME)

RVP   Reid vapour pressure

SHF   separate hydrolysis and fermentation (technology for cellulosic ethanol production)

SME   soy methyl ester (a type of FAME)

SO_x oxides of sulphur

SSF   simultaneous saccharification and fermentation (technology for producing cellulosic ethanol)

THC   total hydrocarbons

USDA   US Department of Agriculture

VOCs   volatile organic compounds

WEO   *Word Energy Outlook*, IEA publication

WTO   World Trade Organization

REFERENCES

AAA, 2002, AAA Response to the Environment Australia Issues Paper (see EIAP, 2001), "Setting the Ethanol Limit in Petrol", http://www.aaa.asn.au/submis/2002/Ethanol_March02.pdf

ADEME, 2003, Journée Débat Biocarburants, Dossier de Presse, http://www.ademe.fr/presse/communiques/CP-2003-04-08.htm

ADM, 1997, Running Line-Haul Trucks on Ethanol: The Archer Daniels Midland Experience, US Department of Energy, November.

AGRI, 2002, "Production and Use of Ethanol from Agricultural Crops", Agricultural Research Institute Kromeriz, Department of Grain Quality, http://www.vukrom.cz/www/english/quality/ukol.htm

Agterberg, A.E. and A.P.C. Faaij, undated, "Biotrade: International Trade in Renewable Energy from Biomass", Department of Science, Technology and Society, Utrecht University, Netherlands.

Ahmed, I., 2001, "Oxygenated Diesel: Emissions and Performance Characteristics of Ethanol-Diesel Blends in CI Engines", Pure Energy Corporation, http://www.pure-energy.com/sae2001-01-2475.pdf

Andress, D., 2001, "Air Quality and GHG Emissions Associated With Using Ethanol in Gasoline Blends", Prepared for Oak Ridge National Laboratory, UT-Battelle, Inc.

ANL, undated, "Using Ethanol as a Vehicle Fuel", Argonne National Laboratory, Transportation Technology R&D Centre, http://www.transportation.anl.gov/ttrdc/ competitions/ethanol_challenge/ethanol.html

ARS, 2003, USDA, Agricultural Research Service, Better Cold-Weather Starts for Biodiesel Fuel, Press Release, 2003, http://www.ars.usda.gov/is/AR/archive/apr98/cold0498.htm

BAA, 2003a, "Fact Sheet: Biodiesel Usage", Biodiesel Association of Australia, http://www.biodiesel.org.au/

BAA, 2003b, "Fact Sheet: Biodiesel Performance", Biodiesel Association of Australia, http://www.biodiesel.org.au/

Baumert, K., 1988, "Carbon Taxes vs. Emissions Trading: What's the Difference, and Which is Better?" April, http://www.globalpolicy.org/socecon/glotax/carbon/ct_et.htm

BBI, 2003, "Peru Plan Aims to Supply California with Ethanol Fuel", http://www.bbiethanol.com/news/view.cgi?article=724

Beard, L.K., 2001, "The Pluses and Minuses of Ethanol and Alkylates for Gasoline Blending: A Carmaker's Perspective", Presentation to Lawrence Livermore Workshop, San Francisco, CA, April 10-11, http://www-erd.llnl.gov/ethanol/proceed/autoupd.pdf

BEI, 2002, "Fuel Alcohol Programme", Brazilian Embassy to India, http://www.indiaconsulate.org.br/comercial/p_exportadores_indianos/ethanol.htm

Cadu, J., 2003, data on ethanol yields from his unpublished review of studies, personal communication.

Canada, 2003, Ethanol Expansion Program News Release, http://www.climatechange.gc.ca/english/publications/announcement/news_release1.html

Canada MoA, 2003, "Fact Sheet 2003" Canada Ministry of Agriculture and Agri-foods, National Biomass Ethanol Program, http://www.agr.gc.ca/progser/nbep_e.phtml

C&T Brazil, 2002, "The Cost of Ethanol", http://www.mct.gov.br/clima/ingles/comunic_old/alcohol4.htm

CGB, 2003, "Biocarburants: Pourquoi le pétrole vert séduit-il aujourd'hui l'Europe?", Confédération Générale des Planteurs de Betteraves, www.cgb-france.fr

Coltrain, D., 2002, "Biodiesel: Is it Worth Considering?", Presentation to 2002 Risk and Profit Conference, Kansas, Aug 15-16, http://www.agmanager.info/agribus/energy/Risk%20&%20Profit%208-02.pdf

CONCAWE, 2002, "Energy and Greenhouse Gas Balance of Biofuels for Europe – an Update", prepared by the CONCAWE ad-hoc group on alternative fuels, Brussels, April.

CRFA, 2003, "Emissions Impact of Ethanol", Canadian Renewable Fuels Association, http://www.greenfuels.org/emissionimpact.html

CSU, 2001, "Alcohol for Motor Fuels", Colorado State University, http://www.ext.colostate.edu/pubs/farmmgt/05010.html

DA, 2002, "Infrastructure Requirements for an Expanded Fuel Ethanol Industry", Downstream Alternatives Inc., Phase II Project Deliverable Report, Oak Ridge National Laboratory Ethanol Project.

DA, 2000, "The Current Fuel Ethanol Industry: Transportation, Marketing, Distribution, and Technical Considerations", Downstream Alternatives Inc., for USDOE, http://www.afdc.doe.gov/pdfs/4788.pdf

DA, 1999, "The Use of Ethanol in California Clean-burning Gasoline; Ethanol Supply/Demand and Logistics", Downstream Alternatives Inc., prepared for the Renewable Fuels Association.

DATAGRO, 2002, Controle de Frota E Consumo Especificio, September.

Defra, 2003, "Common Agricultural Policy (CAP) Reform: Summary of Agreement of 26 June 2003", http://www.defra.gov.uk/farm/capreform/agreement-summary.htm

DESC, 2001, "Fast Pyrolysis of Bagasse to Produce BioOil Fuel for Power Generation", DynaMotive Energy Systems Corp., www.dynamotive.com/biooil/technicalpapers/2001sugarconferencepaper.pdf

Delucchi, M.A., 2004, personal communication. Some related information available in the series of publications at http://www.its.ucdavis.edu/pubs/pub2003.htm

Delucchi, M.A., 2003, "A Lifecycle Emissions Model (LEM): Lifecycle Emissions from Transportation Fuels, Motor Vehicles, Transportation Modes, Electricity Use, Heating and Cooking Fuels, and Materials: Documentation of Methods and Data", UCD-ITS-RR-03-17, Institute of Transportation Studies, U.C. Davis, http://www.its.ucdavis.edu/faculty/delucchi.htm

Delucchi, M.A., 1993, *Emissions of Greenhouse Gases from Transportation Fuels and Electricity*, ANL/ESD/TM-22, published by Argonne National Laboratory.

Dinus, R.J., 2000, "Genetic Modification of Short Rotation Poplar Biomass Feedstock for Efficient Conversion to Ethanol", prepared for the Bioenergy Feedstock Development Program, Environmental Sciences Division, Oak Ridge National Laboratory, http://bioenergy.ornl.gov/reports/dinus/

DiPardo, J., 2002, "Outlook for Biomass Ethanol Production and Demand", U.S. Energy Information Administration. Washington DC.

DOE, 2002a, "FY 2003 Budget in Brief", Office of Energy Efficiency and Renewable Energy, http://www.eren.doe.gov/budget/pdfs/fy03_summary.pdf

DOE, 2002b, Biorefinery Project Awards, U.S. Dept. of Energy, Office of Transportation Technologies, http://www.ott.doe.gov/biofuels/whats_new_ archive.html

DOE, 1999, "Biofuels, A Solution for Climate Change", U.S. Dept. of Energy, Office of Transportation Technologies, DOE/GO-10098-580.

DOE, 1998, "Cellulosic Ethanol: R&D Status & Carbon Emissions", U.S. Dept. of Energy.

EAIP, 2001, "Setting the Ethanol Limit in Petrol", Environment Australia Issues Paper, January.

EC, 1998, "Biodiesel in heavy-duty vehicles in Norway – Strategic plan and vehicle fleet experiments", Final report from European Commission ALTENER-project XVII/4.1030/Z/209/96/NORB1, Rapport 18/98.

EC-DG/Ag, 2002, "Prospects for Agricultural Markets 2002-2009", European Commission, Directorate-General for Agriculture.

EC-DG/Ag, 2001, "Agriculture in the European Union - Statistical and Economic Information 2001", EC, Directorate-General for Agriculture.

EC-JRC, 2002, "Techno-economic Analysis of Bio-alcohol Production in the EU: a Short Summary for Decision-makers", Institute for Prospective Technological Studies, Seville, April.

EESI, 2003, *Ethanol Climate Protection Oil Reduction: A Public Forum*, Vol. 3, Issue II, Interview with Jeff Passmore, Executive Vice President of Iogen Corp, April 2003, http://www.eesi.org/publications/Newsletters/ECO/eco% 2019.PDF

EF, 2002, "Argentina, Bio-diesel and the CDM", *Environmental Finance*, February, http://www.sagpya.mecon.gov.ar/00/index/biodisel/Articulo %20con%20Gaioli%20en%20Environ.%20Finance.pdf

EPA, 2003, "Fuel Economy Impact Analysis of RFG", U.S. Environmental Protection Agency, http://www.epa.gov/orcdizux/rfgecon.htm

EPA, 2002a, "Clean Alternative Fuels: Ethanol Fact Sheet", U.S. Environmental Protection Agency.

EPA, 2002b, "A Comprehensive Analysis of Biodiesel Impacts on Exhaust Emissions", EPA420-P-02-001, October.

Ericsson, K. and L. J. Nilsson, 2003, "International Biofuel Trade – A study of Swedish Import", *Biomass and Bioenergy*, forthcoming.

EU, 2001, Proposal for a Directive of the European Parliament and of the Council, "On Alternative Fuels for Road Transportation, and on a Set of Measures to Promote the Use of Biofuels", Brussels, COM(2001) 547 provisional version.

EU-DGRD, 2001, "Bioethanol Added to Fuel", STOA – Scientific and Technological Options Assessment, Briefing Note N° 07/2001 EN, PE nr. 297.566, European Union, Directorate-General for Research, February, http://www.europarl.eu.int/stoa/publi/pdf/briefings/07_en.pdf

Fergusson, M., 2001, Analysis for PIU on Transport in the Energy Review, for the Institute for European Environmental Policy, Final Report, December.

Ferraz, C. and R. Seroa da Motta, 2000, "Automobile Pollution Control in Brazil", Working Paper No 29, jointly published by the International Institute for Environment and Development, London and the Institute for Environmental Studies, Amsterdam. June, http://www.iied.org/docs/enveco/creed29e.pdf

Fischer, G. and L. Schrattenholzer, 2001, "Global Bioenergy Potential Through 2050", *Biomass and Bioenergy*, Vol. 20.

F.O. Lichts, 2004, "World Ethanol and Fuels Report", 10 February.

F.O. Lichts, 2003, "World Ethanol and Fuels Report", 26 October and supplementary data provided by F.O. Lichts.

Ford, 2003, "Alternative Fuel Vehicles: Ethanol", Ford Motor Company, http://www.ethanol.org/pdf/ford_ffv.pdf

Forum, 2000, Forum on Ethanol Blending in Gasoline in the Northeast and Mid Atlantic, June 27, http://www.nrbp.org/pdfs/pub23.pdf

France, Assemblée Nationale, 2000, "Les Biocarburants dans l'Union Européenne", Rapport d'Information, déposé par la Délégation pour l'Union Européenne.

FURJ, 1998, "Overview of Latin American Technology Development for Avoiding Greenhouse Gas Emissions", Federal University of Rio de Janeiro, November. file paper E10.

Gielen, D.J. *et al.*, 2001, "Biomass for Energy or Materials? A Western European Systems Engineering Perspective", *Energy Policy*, Vol. 29.

GM *et al.*, 2002, "GM Well-to-Wheel Analysis of Energy Use and Greenhouse Gas Emissions of Advanced Fuel/Vehicle Systems – A European Study; ANNEX – Full Background Report", www.lbst.de/gm-wtw

GM/ANL *et al.*, 2001, "Well-to-Wheel Energy Use and Greenhouse Gas Emissions of Advanced Fuel/Vehicle Systems – North American Analysis", Vols 1-3, http://www.transportation.anl.gov/pdfs/TA/163.pdf

GSI, 2000, "A New Beginning In Waste Treatment, Ethanol Derived from Municipal Solid Waste", GeneSyst International, Inc., http://www.genesyst.com/Ethanol_from_Waste/Ethanol_From_MSW.htm

Halvorsen, K.C., 1998, "The Necessary Components of a Dedicated Ethanol Vehicle", December, http://www.westbioenergy.org/reports/55019/55019_final.htm

Hamelinck, C. *et al.*, 2003, "Prospects for ethanol from lignocellulosic biomass: techno-economic performance as development progresses", Utrecht University, Report NWS-E-2003-55, www.cem.uu.nl/nws/www/publica/e2003-55.pdf

Hennepin, 1998, Hennepin County's Experience with Heavy-Duty Ethanol Vehicles, National Renewable Energy Laboratory, January.

Ho, S.P., 1989, "Global Warming Impact of Ethanol v. Gasoline", Presentation at Conference *Clean Air Issues and America's Motor Fuel Business*, Washington DC, October.

Hoogwijk, M. A. *et al.*, 2003, "Exploration of the Ranges of the Global Potential of Biomass for Energy", accepted for publication in *Biomass and Bioenergy*.

IBS, 2003, "Short crop may sweeten the deal", *Indian Business Standard*, 10 December, http://www.business-standard.com/today/story.asp?Menu=22&story=28979

ICGA, 2003, "Questions and Answers about E85 and Flexible Fuel Vehicles", Iowa Corn Growers' Association, http://www.iowacorn.org/

IEA, 2002, *Bus Systems for the Future: Achieving Sustainable Mobility Worldwide*, International Energy Agency, Paris.

IEA/WEO 2002, *World Energy Outlook 2002*, International Energy Agency, Paris.

IEA, 2001, *Saving Oil and Reducing CO_2 Emissions in Transport*, International Energy Agency, Paris.

IEA, 2000a, "Liquid Fuels from Biomass: North America; Impact of Non-Technical Barriers on Implementation", prepared for the IEA Bioenergy Implementing Agreement, Task 27 Final Report, by (S&T)2 Consultants Inc.

IEA, 2000b, "Bioethanol in France and Spain: Final Report", prepared for the IEA Bioenergy Implementing Agreement, Task 27 Final Report, by V. Monier and B. Lanneree, Battelle Pacific Northwest Laboratories and Taylor Nelson Sofres Consulting.

IEA, 2000c, "Alternative fuels (ethanol) in Sweden" prepared for IEA Bioenergy Implementing Agreement, Task 27 Final Report, by N. Elam, Atrax Energi, AB.

IEA, 2000d, "Biodiesel in Europe: Systems Analysis, Non-technical Barriers", prepared for the IEA Bioenergy Implementing Agreement, Task 27 Final Report, by F. Eibensteiner and H. Danner, SySan, IFA.

IEA, 1999, *Automotive Fuels for the Future: The Search for Alternatives*, OECD/IEA, Paris.

IEA, 1994, *Biofuels*, OECD/IEA, Paris.

IFP, 2003, "Biofuels from Biomass", Institut Français du Pétrole, Rueil-Malmaison, France.

IPCC, 2001, "Technological and Economic Potential of Greenhouse Gas Emissions Reduction", Third Assessment Report: Mitigation.

IPTS, 2003, "Biofuel Production Potential of EU-Candidate Countries", Institute for Prospective Technological Studies.

ITDG, 2000, "Biogas and Biofuels", Technical Brief, http://www.itdg.org/html/technical_enquiries/docs/biogas_liquid_fuels.pdf

Johnson, F.X., 2002, "Bioenergy from Sugar Cane for Sustainable Development and Climate Mitigation: Options, Impacts and Strategies for International Co-operation", Stockholm Environment institute, Sweden.

Kadam, K.L., 2002, "Environmental Benefits on a Life-cycle Basis Implications of Using Bagasse-Derived Ethanol as a Gasoline Oxygenate in India", *Energy Policy*, Vol. 30: 5, April.

Kadam, K.L., 2000, "Environmental Life Cycle Implications of Using Bagasse-Derived Ethanol as a Gasoline Oxygenate in Mumbai (Bombay)", Prepared for the National Energy Technology Laboratory, Pittsburgh, Pennsylvania, USA and USAID, New Delhi, India, November.

Keeney, D.R. and T.H. DeLuca, 1992, "Biomass as an Energy Source for the Midwestern U.S.", *American Journal of Alternative Agriculture*, Vol. 7.

Kingsman News, 2003, "Brazil takes steps to avoid ethanol shortage", Kingsman Ethanol and Industry News, February 2003.

Klass, D.L., 1998, "Los Angeles Evaluation of Methanol- and Ethanol-Fuelled Buses", http://www.bera1.org/LA-buses.html

Koizumi, T., 2003, "The Brazilian Ethanol Program: Impacts on World Ethanol and Sugar Markets", paper for OECD Workshop on Biomass and Agriculture, Vienna, June 10-13.

Korbitz, W., 2002, "New Trends in Developing Biodiesel World-wide", Austrian Biofuels Institute, Vienna, Austria, presented at "Asia Biofuels" Conference, Singapore, April 22-23, http://www.nachhaltigkeitsrat.de/service/download_e/pdf/Asia_Biofuels_Worldwide.pdf

Laydner, L.O., 2003, "The Search for New Markets", *Brazil Equity Update, Brazilian Fuel Alcohol Sector*, October 6.

Levelton 2000a, "Engineering Assessment of Net Emissions of Greenhouse Gases from Ethanol-Gasoline Blends in Southern Ontario", January 2000, http://www.tc.gc.ca/programs/Environment/climatechange/docs/ETOH-FNL-RPTAug30-1999.htm

Levelton, 2000b, "Assessment of Net Emissions of Greenhouse Gases from Ethanol-Blended Gasolines in Canada: Lignocellulosic Feedstock", Levelton Engineering Ltd, R-2000-2, in association with(S&T)2 Consultants Inc.

Levelton, 1999, "Alternative and Future Fuels and Energy Sources for Road Vehicles", Levelton Engineering Ltd, prepared for Canadian Transportation Issue Table, National Climate Change Process.

Lif, A., 2002, "Pure Alcohol Fuel and E-diesel Fuel Additives Technologies", AkzoNobel Surface Chemistry, April.

Lightfoot, H.D. and C. Green, 2002, "An Assessment of IPCC Working Group III Findings in Climate Change 2001: Mitigation of the Potential Contribution of Renewable Energies to Atmospheric Carbon Dioxide Stabilization", McGill Centre for Climate and Global Change Research, Canada.

Lloyd, A.C. and T. A. Cackett, 2001, "Diesel Engines: Environmental Impact and Control", *Journal of Air and Waste Management,* Vol. 51, June.

Lockart, M., 2002, Growmark Corporation Press Release, August, http://www.agandenvironment.com/news/news_20020828.htm

Lorenz, D. and D. Morris, 1995, "How Much Energy Does it Take to Make a Gallon of Ethanol?", Institute for Local Self-reliance, Washington DC, August.

Lynd, L., 2004, Personal communication concerning global biofuels potential.

Lynd, L. *et al.*, 2003, "Bioenergy: Background, Potential, and Policy", policy briefing prepared for the Center for Strategic and International Studies, submitted.

Macedo, I.C. *et al.*, 2003, "Greenhouse Gas Emissions in the Production and Use of Ethanol in Brazil: Present Situation (2002)", prepared for the Secretariat of the Environment, São Paulo.

Macedo, I.C., 2001, "Converting Biomass to Liquid Fuels: Making Ethanol from Sugar Cane in Brazil", Part III, Chapter 10 of "Energy as an Instrument for Socio-Economic Development ", UN Development Programme, http://www.undp.org/seed/energy/policy/ch10.htm

Maniatis, K. and E. Millich, 1998, "Energy from Biomass and Waste: The Contribution of Utility-Scale Biomass Gasification Plants", *Biomass and Bioenergy*, Vol. 15: 3.

Marland, G. and A. Turhollow, 1990, "CO_2 Emissions from the Production and Combustion of Fuel Ethanol from Corn", Oak Ridge National Laboratory, Environmental Sciences Div., prepared for US Department of Energy.

MBEP, 2002, "Fact Sheet", Michigan Biomass Energy Program, http://www.michiganbioenergy.org/ethanol/edieselfacts.htm

McCormick, R.L. and R. Parish, 2001, "Advanced Petroleum Based Fuels Program and Renewable Diesel Program Milestone Report: Technical Barriers to the Use of Ethanol in Diesel Fuel", National Renewable Energy Laboratory, November.

Midwest, 1994, "Biodiesel Cetane Number Engine Testing Comparison to Calculated Cetane Index Number", www.biodiesel.org

Mittelbach, M. 2002, "Experience with Biodiesel from Used Frying Oil in Austria", Karl-Franzens-University Graz, October, http://www.portalenergy.com/balpyo/renew14/04.pdf

Moreira, J.R., 2003, personal contact with Prof. José Roberto Moreira, Brazilian Biomass Reference Centre – CENBIO – Rua Francisco Dias Velho.

Moreira, J.R., 2002, "Can Renewable Energy Make Important Contribution to GHG Atmospheric Stabilization?", LAMNET Third Project Workshop, December.

Murray, L.D., 2002, *Avian Response to Harvesting Switchgrass in Southern Iowa*, Thesis, Iowa State University.

Murthy, B.S., 2001, "Alcohols in Diesel Engines", SAE India, http://www.saeindia.org/home/alcoholindieselengines.html

MSU, 1999, "Use of Mid-Range Ethanol/Gasoline Blends in Unmodified Passenger Cars and Light Duty Trucks", Kirk Ready *et al.*, Minnesota Center for Automotive Research, Minnesota State University, 1999.

NBB, undated, "Facts about Biodiesel", National Biodiesel Board (US), http://www.biodiesel.org/markets/gen/default.asp

Neff, T.L., 1997, Improving Energy Security in Pacific Asia: Diversification and Risk Reduction for Fossil and Nuclear Fuels, Center for International Studies, Massachusetts Institute of Technology, December 1997, http://www.nautilus.org/papers/energy/NeffPARES.pdf

NEVC, 2002, *National Ethanol Vehicle Coalition Newsletter*, Vol. 7: 9 April, http://www.e85fuel.com/news/043002fyi.htm

Novem, 2003, "BIO-H$_2$: Application of Biomass-related Hydrogen Production Technologies to the Dutch Energy Infrastructure, 2020-2050", http://www.novem.nl/default.asp?menuId= 10&documentId=26620

Novem/Ecofys, 2003, "Biofuels in the Dutch Market: A Fact-finding Study", prepared by Ecofys for Novem, GAVE analysis programme, project \# 2GAVE-03.12.

Novem/ADL, 1999, *Analysis and Evaluation of GAVE Chains*, Vol. 1-3, GAVE analysis programme, http://www.novem.nl/default.asp?menuId=10 &documentId=699

NRC, 1999, "Review of the Research Strategy for Biomass-Derived Transportation Fuels", U.S. National Research Council, Commission on Engineering and Technical Systems.

NREL, 2002, *Handbook for Handling, Storing, and Dispensing E85*, Produced for the U.S. Department of Energy (DOE) by the National Renewable Energy Laboratory, DOE/GO-1002001-956, April.

NREL, 2001, "Biodiesel: Handling and Use Guidelines", National Renewable Energy Laboratory, NREL/TP-580-30004. http://www.ott.doe.gov/biofuels/pdfs/biodiesel_handling.pdf

NREL, 2000, "Biodiesel: the Clean, Green Fuel for Diesel Engines", Produced for the U.S. Department of Energy (DOE) by the National Renewable Energy Laboratory, DOE/GO-102000-1048, May, http://www.ott.doe.gov/biofuels/environment.html

NREL, 1998, "Final Results from the State of Ohio Ethanol-Fuelled Light-Duty Fleet Deployment Project", National Renewable Energy Laboratory, August.

Nylund, N.-O., 2000, "Fuel Trends in Europe", VTT Energy, presented at Windsor Workshop on Transportation Fuels, Toronto, http://www.windsorworkshop.ca/downloads/Nylund.pdf

OECD, 2003, Environmental Goods: "A Comparison of the APEC and OECD Lists" Environment and Trade Directorates, http://www.olis.oecd.org/olis/2003doc.nsf/LinkTo/com-env-td(2003)10-final

ODE, 2003, "Biomass Energy: Cost of Production" Oregon Department of Energy, http://www.energy.state.or.us/biomass/Cost.htm

ORNL, 2000, "Air Quality and GHG Emissions Associated With Using Ethanol in Gasoline Blends", Oak Ridge National Laboratory.

Otte C. *et al.*, 2000, "The Newest Silverado: A Production Feasible Ethanol (E85) Conversion by the University of Nebraska-Lincoln", The University of Nebraska-Lincoln Department of Mechanical Engineering and SAE, Inc.

PA, 2001, "Congressional Committee Holds Hearing on MTBE Concerns", Pennsylvania Dept. of Environmental Protection, http://www.dep.state.pa.us/dep/deputate/polycomm/update/11-02-01/1102014948.htm

Peelle, E., 2000, "Stakeholder Views and Concerns about Bioenergy: Organizational Focus, Driver Issues and Uncertainty", Oak Ridge National Laboratory, http://bioenergy.ornl.gov/papers/bioen00/peele.html

Pimentel, D., 2001, "The Limits of Biomass Energy", *Encyclopaedia of Physical Sciences and Technology*, September.

Pimentel, D., 1991, "Ethanol Fuels: Energy Security, Economics, and the Environment", *Journal of Agricultural and Environmental Ethics*, Vol. 4.

Rae, A., 2002, "E-Diesel: An Immediate and Practical Air Quality and Energy Security Solution", BAQ 2002 Conference, Hong Kong, December, http://www.cse.polyu.edu.hk/~activi/BAQ2002/BAQ2002_files/Proceedings/PosterSession/53.pdf

Ragazzi, R. and K. Nelson, 1999, "Evaluation of the Use of 10% Ethanol", Colorado State U. for Colorado Dept. of Public Health and Environment, http://www.cdphe.state.co.us/ap/down/oxyfuelstudy.pdf

Raneses, A.R. *et al.*, 1999, "Potential Biodiesel Markets and their Economic Effects on the Agricultural Sector of the US", *Industrial Crops and Products*, Vol. 9.

Read, P., 2003, "The Policy, Economic, Environmental, and Social Aspects of Linking Bioenergy with Carbon Storage in a Sequential Decision Approach to the Threat of Abrupt Climate Change", OECD Workshop on Biomass and Agriculture, 10-13 June, Vienna.

RESP, 2003a, "Blue Buses Pave the Way to Greener Streets", Response Online, http://www.responseonline.com/archi/etha.htm

RESP, 2003b, "Thumbs up for E-diesel, Truck trials in Denmark Confirm Benefits to Public Health", http://www.responseonline.com/thumb.htm

Reuters, 2001, "Japan Eyes Ethanol to Cut Greenhouse Gas Emissions", December 20, http://www.bbiethanol.com/news/view.cgi?article=401

RFA, 1999, "Ethanol Industry Outlook: 1999 and Beyond", Renewable Fuels Association.

Ribeiro, S.K., 2000, "The Brazilian Fuel Alcohol Programme", Case Study No. 8, in Chapter 16 of IPCC, May 2000, *Methodological and Technological issues in Technology Transfer*, http://www.grida.no/climate/ipcc/tectran/336.htm

Rice B. *et al.*, 1997, "Bio-diesel Production based on Waste Cooking Oil: Promotion of the Establishment of an Industry in Ireland", Teagasc, http://www.biodiesel.org/resources/reportsdatabase/reports/gen/19970 901_gen-190.pdf

RMI, 2003, "U.S. Energy Security Facts For a Typical Year", 2000, Rocky Mountain Institute, June, http://www.rmi.org/images/other/S-USESFsheet.pdf

SAE, 2001, "National Ethanol Vehicle Challenge Design Competition", http://www.saeindia.org/home/NEVC.htm

Scharmer, K., 2001, "Biodiesel: Energy and Environmental Evaluation, RME", Union Zur Förderung von Oel und Proteinpflanzen E.V.

Schindler, J. and W. Weindorf, 2000, "Fuels for Transportation Derived from Renewable Energy Sources", Hyforum 2000, Munich, www.hyweb.de/Wissen/pdf/hyforum2000.pdf

Schremp, G., 2001, "California Issues – Expanded Use of Ethanol and Alkylates", LLNL Workshop, Oakland CA, April 10-11, http://www-erd.llnl.gov/ethanol/proceed/cecupd.pdf

Schremp, G., 2002, "Status for California's plans for MTBE and the Federal Oxygenate Waver Request", World Fuels Conference, San Antonia Tx, March 21.

SEDF, 1997, "Ethanol as Transport Fuel in Sweden", Swedish Ethanol Development Foundation, http://www.nf-2000.org/secure/Other/F447.htm

Shapouri, H., J. Duffield and M. Wang, 2002, "The Energy Balance of Corn Ethanol: An Update", US Dept. of Agriculture, Agriculture Economic Report #813, http://www.transportation.anl.gov/pdfs/AF/265.pdf

Shapouri, H. *et al.*, 1995, "Estimating the Net Energy Balance of Corn Ethanol", U.S. Dept. of Agriculture, Economic Research Service, AER-721.

Sheehan, J. 2000. "Feedstock Availability and the Role of Bioethanol in Climate Change", National Renewable Energy Laboratory, Presented at 13th International Alcohol Fuels Symposium, Stockholm, Sweden, July.

Silva, G. *et al.*, 1978, "Energy Balance for Ethyl Alcohol Production from Crops", *Science,* Vol. 201: 4359, September.

Sims, R., 2003, "The Triple Bottom Line Benefits of Bioenergy for the Community", for OECD Workshop on Biomass and Agriculture, Vienna, Austria, 10-13 June.

Sreenath, H.K. *et al.*, 2001, "Ethanol Production from Alfalfa Fiber Fractions by Saccharification and Fermentation", *Process Biochemistry*, Vol. 36, pp. 1199-1204, http://www.fpl.fs.fed.us/documnts/pdf2001/sreen01a.pdf

SRI, undated, "Efficiency Improvements Associated with Ethanol-Fuelled Spark-Ignition Engines", Southwest Research Institute, http://www.swri.edu/4org/d03/engres/spkeng/sprkign/pbeffimp.htm

Tickell, J., 2000, *From the Fryer to the Fuel Tank: The Complete Guide to Using Vegetable Oil as an Alternative Fuel.*, Veggie Van Publications.

Ugarte, D. and M. Walsh, 2002, "Synergism between Agricultural and Energy Policy: The Case of Dedicated Bioenergy Crops", Agricultural Policy Analysis Center, University of Tennessee, http://agpolicy.org/ppap/pdf/02/biocrop.pdf

UK DVLA, 2004, "Rates of Vehicle Excise Duty", U.K. Driver and Vehicle Licensing Agency, http://www.dvla.gov.uk/vehicles/taxation.htm

Ullmann, J. and R. Bosch, 2002, "The Influence of Biodiesel Properties on Fuel Injection Equipments", Presentation to the Seminario Internacional de Biodiesel Curitiba, 24-26 October.

UNFCCC, 2003, "Measures to Mitigate Emissions of Greenhouse Gases", The UN Framework Convention on Climate Change, The Czech Republic's Third National Communication, http://unfccc.int/resource/docs/natc/pam/czepamn3.pdf

US Census, 2001, "U.S. International Trade in Goods and Services", Foreign Trade Statistics, November 2001.

USDA, 2003a, "Track Records, US Crop Production", U.S. Dept. of Agriculture, National Agricultural Statistics Service (database), http://www.usda.gov/ nass/pubs/trackrec/track02a.htm\#top.

USDA, 2003b, "Brazilian Sugar", U.S. Dept. of Agriculture, October, http://www.fas.usda.gov/htp/sugar/2003/ Brazilsugar03.pdf

USDA, 2002, "Agricultural Baseline Projections to 2011", Office of the Chief Economist, U.S. Department of Agriculture, Prepared by the Interagency Agricultural Projections Committee, Staff Report WAOB-2002-1.

USGS, 2003, "Water Resources of Pennsylvania", U.S. Geological Survey, http://wwwpah2o.er.usgs.gov/

van Thuijl, E., *et al.,* "An Overview of Biofuel Technologies, Markets and Policies in Europe", ECN-C-03-008, Netherlands.

Vargas, A., 2002, "Ethanol Potential in Costa Rica", ECOMAP Unit – International Center of Economic Policy (CINPE-UNA), 3rd LAMNET Workshop – Brazil, December.

Walsh, M.E. *et al.,* 2002, "The Economic Impacts of Bioenergy Crop Production on US Agriculture", ORNL Bioenergy Information Network, http://bioenergy. ornl.gov/papers/wagin/

Walsh, M.E. *et al.,* 2000, "Biomass Feedstock Availability in the United States: 1999 State Level Analysis", ORNL, http://bioenergy.ornl.gov/resourcedata/ index.html

Wang, M., 2003a, "Fuel-Cycle Energy and Greenhouse Emission Impacts of Fuel Ethanol", Centre for Transportation Research, Argonne National Laboratory.

Wang, M. *et al.,* 2003b, "Fuel-Cycle Energy and Emission Impacts of Ethanol-Diesel Blends in Urban Buses and Farming Tractors", Centre for Transportation Research Argonne National Laboratory, July 2003.

Wang, M., 2001a, "Greet Model version 1.5a", revision June 2001 (calculations made by IEA for reference case using the downloadable model, in consultation with author), http://greet.anl.gov

Wang, M., 2001b "Development and Use of Greet 1.6 Fuel-Cycle Model for Transportation Fuels and Vehicle Technologies", Argonne National Laboratory, ANL/ESD/TM-163.

Whims, J., 2002, "Corn Based Ethanol Costs and Margins", Attachment 1, AGMRC, Kansas State U., http://www.agmrc.org/corn/info/ksueth1.pdf

Wiltsee, G., 1998, "Urban Waste Grease Resource Assessment", Appel Consultants, Inc., for National Renewable Energy Laboratory, November, http://www.ott.doe.gov/biofuels/pdfs/urban_waste.pdf

Yamamoto, H. *et al.*, 2001, "Evaluation of Bioenergy Potential with a Multi-regional Global-land-use-and-energy Model", *Biomass and Bioenergy*, Vol. 21.

The Online Bookshop

International Energy Agency

All IEA publications can be bought
online on the IEA website:

www.iea.org/books

You can also obtain PDFs of all IEA books
at 20% discount.

Books published in 2002 and before
can be downloaded in PDF, free of charge,
on the IEA website.

IEA BOOKS

Tel: +33 (0)1 40 57 66 90
Fax: +33 (0)1 40 57 67 75
E-mail: books@iea.org

International Energy Agency
9, rue de la Fédération
75739 Paris Cedex 15, France

International Energy Agency, 9 rue de la Fédération, 75739 Paris Cedex 15

PRINTED IN FRANCE BY CHIRAT
61 2004 16 1P1 ISBN 92 64 01 51 24   April 2004